# Empowering Women in STEM

Women in STEM are constantly facing new challenges every day. By sharing their stories and ways in which they have and continue to overcome these hurdles, they can help others find the strength to persevere and succeed in these fields. This mix of authors from varying backgrounds all share the same passion – to encourage more women into STEM – and they have placed their proverbial hearts on their sleeves and documented their journeys to assist readers to either enter or stay within STEM fields.

*Empowering Women in STEM: Personal Stories and Career Journeys from Around the World* discusses the application process of root cause analysis and ways to introduce STEM to other generations. It offers an insider view of armed forces and allows readers to gain more perspective on ways male advocates can help female colleagues in STEM fields. It includes a father's perspective on change within the engineering industry, how he's mentoring new female engineers, and ways to help them as they evolve. The book captures firsthand accounts of STEM professionals in various fields as they recount experiences that have helped them to navigate their career paths. This book also demonstrates how life doesn't have to follow the timelines proposed by society and how females can become CEOs and command other top-level positions in engineering companies.

In addition to having women from across the globe share their stories about various fields, you will hear from both military and civilian male advocates who share ways to empower others within the industry. This book is written for professionals who may be considering a switch of career or deciding to leave STEM. It is also useful to university students who are trying to figure out their career choices and paths and gain more insight into possible career opportunities in STEM.

# Empowering Women in STEM

## Personal Stories and Career Journeys from Around the World

Edited by
Sanya Mathura

CRC Press is an imprint of the
Taylor & Francis Group, an **informa** business

First edition published 2023
by CRC Press
6000 Broken Sound Parkway NW, Suite 300, Boca Raton, FL 33487-2742

and by CRC Press
4 Park Square, Milton Park, Abingdon, Oxon, OX14 4RN

*CRC Press is an imprint of Taylor & Francis Group, LLC*



*Library of Congress Cataloging-in-Publication Data*
Names: Mathura, Sanya, editor.
Title: Empowering women in STEM : personal stories and career journeys from around the world / edited by Sanya Mathura.
Other titles: Empowering women in science, techlology, engineering, and mathematics
Description: First edition. | Boca Raton : CRC Press, 2023. | Includes bibliographical references and index.
Identifiers: LCCN 2022034130 (print) | LCCN 2022034131 (ebook) | ISBN 9781032372648 (pbk) | ISBN 9781032373317 (hbk) | ISBN 9781003336495 (ebk)
Subjects: LCSH: Women in engineering. | Women in science. | Women in mathematics. | Women--Professional relationships. | Engineering--Vocational guidance. | Science--Vocational guidance. | Mathematics--Vocational guidance.
Classification: LCC TA157.5 .E46 2023 (print) | LCC TA157.5 (ebook) | DDC 620.0082--dc23/eng/20221006
LC record available at https://lccn.loc.gov/2022034130
LC ebook record available at https://lccn.loc.gov/2022034131

ISBN: 978-1-032-37331-7 (hbk)
ISBN: 978-1-032-37264-8 (pbk)
ISBN: 978-1-003-33649-5 (ebk)

DOI: 10.1201/9781003336495

Typeset in Times
by SPi Technologies India Pvt Ltd (Straive)

# Dedication

One of the best allies any female in STEM can have is her father. Fathers influence and guide their unique creations from conception. They do so through unconditional love and can provide guidance for any pathway their daughter may choose. Fathers nurture their daughters' inquisitive natures and create an environment where they can freely explore their talents without fear of failure or judgement. Most fathers help lay the foundation for women who change the world.

I dedicate this book to my father, Ambrose Mathura. May you rest in peace and may the angels welcome you with the same open arms and loving heart as your own. Thank you for believing in me, supporting me, and nurturing me to become the person I am today. You were a teacher, an active community member and a doting father able to understand your assignment in life early. You understood that the formative years of childhood and adolescence are critical to the direction of not only your students, but your own children's life paths.

Thank you for respecting and making space for me to grow and develop, for allowing me to pursue my dreams to be an engineer, and most importantly for refusing to give up on me. I am forever grateful for the pillars of humility, unconditional love, and faith that you instilled in me.

You have empowered me to break the proverbial glass ceilings in my career and pave the way for the future generation of women in STEM. I will continue to encourage these women to be fearless in the pursuit of their dreams, **to be themselves unapologetically**, and to never forget to pay it forward. Sleep with the angels until we meet again.

# Contents

# Preface

In today's society, we have made significant changes towards the acceptance of women in traditionally male-dominated roles especially within the STEM fields. Though these changes are not as prevalent as one would expect, there is an entire generation of women who have faced these challenges and triumphed. These women are paving the way forward for the new generation while inspiring those who may be thinking about leaving these industries to reconsider. This book recounts some of the challenges these women have overcome and provides readers with some words of wisdom for approaching particular situations.

Women are not in this fight alone; our male allies have continuously supported us in bridging the gaps and overcoming the obstacles faced and this book tells no different story. There are stories in here which speak to overcoming imposter syndrome, ways to support your daughter/s or female co-workers, and of course understanding the power of collaboration. Our male authors have very diverse backgrounds, but there is one unifying factor: their support for women in these traditionally male-dominant fields.

In this book, the term STEM represents Science, Technology, Engineering, and Mathematics. Recently, this acronym has evolved into STEAM, which now includes Arts. Although this "new letter" has been now been added to the STEM acronym, it has always existed but is now being recognized. Creativity is what defines humanity. Thus, without the arts, we are simply writing a program (as put by one of our contributors). Throughout this book, the terms STEM and STEAM may be interchangeably used.

This book is dedicated to uncovering the personal stories and career journeys of some women in STEM. It helps us understand that the pathway forward for most women in STEM may not always be filled with rainbows and butterflies. We uncover the topics of taking a nontraditional route to becoming a female CEO in power engineering and the collision course experienced by a female research engineer in developing crash tests and identifying road safety issues. We also take a look at the dangers of boxing yourself in by casual branding and how to overcome pitfalls like these.

Our male advocates show us how root cause analysis (RCA) can be applied to any STEM discipline and the way it was introduced to a 10-year-old who analyzed bullying in her class. We move from the traditional classroom to the military classroom where the importance of having male advocates is thoroughly explored and the concept of imposter syndrome is demolished. The question of gender in STEM is examined from the vantage point of a veteran engineer in the UK who is trying to mentor younger engineers and pave a way forward for his daughter.

While infertility may still seem to be a taboo topic especially for women within the STEM fields, one of our authors has given us some understanding into ways of navigating your career while handling infertility. She even gives some advice for co-workers and ways they can provide support to women experiencing similar types of situations. Furthermore, we get a first-hand account of the journey of a chemist from Portugal who transitioned into a reliability environment in Qatar. She additionally talks about being pregnant during COVID-19 while in a foreign country.

A female power engineer shares her journey with us as she pioneered and piloted various women in STEM ventures across the globe. She provides readers with her experiences and her role in advocating policies for change. Another group of talented individuals from Anukarniya, comprising of university graduates, fellows, and other researchers, came together to inform us about the challenges women in STEM tend to experience and offer possible solutions. They touch on ways to improve yourself, understand the challenges in the field, and ways to face failures.

The power of collaboration is explored and its many advantages, which are not traditionally taught within the classroom. Diversity of thought is brought into focus to help readers recognize its true potential. Five unconventional life lessons are shared with our readers, as this female engineer shows us her transition from walking the plant to writing. She lets us know about the various career paths that can be taken for engineers and how motherhood helped guide her in her new role.

A university student gives us some insight into the ways in which her abilities charted her progression into various fields. She talks about the impacts of being placed into special classes and the expectations associated with these. Interestingly enough, her father also makes a contribution where he talks about his journey into STEM and how his journey was charted. We have an account from a Nigerian engineer who studied and worked on three different continents. She lets us know more about the culture shocks for her and her journey toward being an advocate for women in STEM.

The significance of planning personal goals while navigating the oil and gas industry is highlighted as well. We follow the journey of a junior engineer who transitions from leading projects and teams to leading strategy in an Edtech start-up. She highlights the soft skills, organizational culture, and the significance of upskilling. Moreover, another journey of a female agricultural engineer who transitioned from service to sales to teaching gives us awareness of the challenges within this industry. She shares ways to impact the younger generation and help them on their own journeys.

Lastly, we have a tribute to five amazing women in STEM who are making huge differences in the world. The author takes us along the journeys of women who have launched programs such as "Introduce a girl to engineering" in the United States, to a female pump engineer from Spain who solves problems in the water industry. This author additionally captures the story of a female reliability engineer who has to break barriers on her journey toward becoming a senior-level engineer. Furthermore, this author describes the journey of a young female mechanical engineer involved in Society for Maintenance and Reliability Professionals and that of another female engineer who redesigns systems to make them more efficient.

It is the hope that this book will encourage more women to pursue STEM-related careers, while also inspiring those within these fields to continue the fight and shatter those glass ceilings for the next generation. This book can also be used as a guide for our male allies who want to assist but are not sure where to start. It will provide others with first-hand experiences of women and men who have faced difficult situations. We need to be each other's keepers in this ongoing battle to ensure that we achieve equality and diversity in every workplace.

# Acknowledgements

I would like to sincerely thank all of the contributors for making the time in their busy schedule to write their chapters. Each contributor has shared snapshots of their life with our readers and worked arduously to document their milestones and advice for all. It is truly a remarkable event where people from across the globe can come together to support a singular cause, *Empowering Women in STEM*.

Thank you to everyone who contributed to this book and shared their lives with us, especially;

Robert J. Latino
Emma Holloway
Heather Eason
Matthew J. Walker
Becky Mueller
Daniel Shorten
Priya Santhanam
Shalini Aggarwal
Beauty Kumari
Harihar Jaishree Subrahmaniam
Subathra Rajendran
Anand Swaroop
Vrinda Nair
Shadrach Stephens
Erin Gutsche
Bralade Koroye-Emenanjo
Jade Thompson
Liana Roopnarine
Alicia Washington
Vanda Franco
Kathy Nelson
Michael D. Holloway
Frances Christopher
Kari Nathan
Michelle Segrest
Jayne Beck
Elena Rodriguez
Rendela Wenzel
Gina Hutto Kittle
Rebekah Macko

# Editor Biography

**Sanya Mathura** is the founder of Strategic Reliability Solutions, Ltd, based in Trinidad & Tobago. She works with global affiliates in the areas of Reliability and Asset Management to bring these specialty niches to her clients. She has a BSc in Electrical & Computer Engineering and an MSc in Engineering Asset Management. Sanya has worked in the lubrication industry for the past several years and assists various industries with lubrication-related issues. Sanya is the first person (and first female) in her country and the Caribbean to attain an ICML MLE (International Council for Machinery Lubrication Machinery Engineer) certification as well as the first female in the world to achieve the ICML Varnish and Deposit Prevention and Removal & Varnish and Deposit Identification and Measurement badges.

# Contributors

**Shalini Aggarwal**
The Indian Institute of Technology
Bombay, India

**Frances Christopher**
SGS
United States of America

**Heather Eason**
Select Power Systems, LLC
United States of America

**Vanda Franco**
Petroleum Technology Company W.L.L.
Qatar

**Erin Gutsche**
Words with Purpose, Inc
Canada

**Emma Holloway**
Colorado State University
United States of America

**Michael D. Holloway**
5th Order Industry, LLC
United States of America

**Bralade Koroye-Emenanjo**
Fortune 100 Operations Leader
United States of America

**Beauty Kumari**
Asian Development Research Institute
India

**Robert J. Latino**
Prelical Solutions, LLC
United States of America

**Becky Mueller**
Insurance Institute for Highway Safety
United States of America

**Vrinda Nair**
Concordia University
Montréal, Quebec, Canada

**Kari Nathan**
Technology Exploration Career Center, East Lewisville ISD
United States of America

**Kathy Nelson**
West Monroe
United States of America

**Subathra Rajendran**
Rapid Serviz
India

**Liana Roopnarine**
GLEAC
United Arab Emirates

**Priya Santhanam**
Amazon
United States of America

**Michelle Segrest**
Navigate Content, Inc
United States of America

**Daniel Shorten**
Optimain Limited
United Kingdom

**Shadrach Stephens**
Re.engineer
United States of America

**Harihar Jaishree Subrahmaniam**
Aarhus University
Denmark

**Anand Swaroop**
White Hat Jr
India

**Jade Thompson**
PCMS Engineering
United Kingdom

**Matthew J. Walker**
Licensed Professional Engineer
United States of America

**Alicia Washington**
Petrochemical M&R Expert
United States of America

# *Section I*

## *STEM Beginnings*

# 1 The Application of an Effective Root Cause Analysis to Any STEM Discipline

*Robert J. Latino*
Prelical Solutions, LLC, United States of America

## CONTENTS



This chapter is going to focus on the 'BASICS' of how to apply an effective root cause analysis (RCA) approach to any undesirable outcome. However, we are sticking to the fundamentals just like blocking and tackling. Everyone has their own way to use such cause-and-effect tree expressions and I am just conveying my preferred one, which is consistent with the PROACT RCA Methodology.

**Defining STEM**

Let's start with the term STEM itself and ensure that we all have a common understanding, before we continue, and everyone is not on the same page. This should also be a life lesson for those interested in STEM-related careers, as effective communication will be critical to your future successes, whether at work or in your personal life. Words matter, because in the absence of a shared understanding, there will most certainly be misinterpretation and undesirable outcomes will be close by.

DOI: 10.1201/9781003336495-2

STEM is a common abbreviation for four (4) closely connected areas of study:

1. **S**cience
2. **T**echnology
3. **E**ngineering and
4. **M**athematics

One common denominator of all these fields of study is that they ALL involve solving problems in order to be successful. To be effective in these fields, we must possess exemplary problem solving, analytical and investigative skills. This is where the concept of root cause analysis, or RCA, comes into play.

## 1.1 MY PERSONAL STORY ABOUT GROWING UP IN A RELIABILITY HOUSEHOLD

Please bear with me for a quick journey down the memory lane, so that you get the context of my upbringing and why this is a personal issue for me. I come from a field called Reliability Engineering. To sum the field up, it is a proactive field of engineering that focuses on ensuring processes are reliable in the future, versus simply fixing things in the present and hoping for they don't recur. Reliability engineers use all sorts of predictive technologies to prioritize risks and identify impending failures, allowing the process to be shut down in a scheduled and safe manner. This avoids the process from failing unexpectedly and costing the organization a lot more money. While reliability engineering is a fairly common and accepted field today, it was not always that way. My father, Charles J. Latino, was a chemical engineer who started his career in the early 1950s. His parents (my grandparents) immigrated to the U.S. and were of very humble means living in New York. However, they were very resilient and proud people, and wanted their children to be successful in their new country. They took great pride in envisioning the 'American Dream' and ensuring their kids did all they could to realize it.

My father worked his own way through college and got a Chemical Engineering degree. He would tell me how he loved math. He reflected on his high school years and he had taken the highest level of math they taught, in his junior year. He prided himself on the fact he had to lobby the school principal to ensure he could take advanced math classes in his senior year … and he did! His first job was at a company in Hopewell, VA called Allied Chemical & Dye Corporation. They made carpet nylon.

I remember my many lunch conversations with Charles and he was always in awe of aviation technologies at the time. He had worked his way up to being the head of Maintenance & Engineering for a plant with around 5000 employees in the mid-60s. He heavily researched the aviation technologies because he wanted to learn how planes could be so much more reliable than his manufacturing site. What could be learned from aviation to make his plant more reliable? As he learned about preventive and predictive technologies, root cause analysis (RCA) and computerized maintenance management systems (CMMS – the equivalent in that day), he started to modify them for application in his plant … and his team's efforts began to make quantum improvements to their overall reliability.

He started to get the attention of Allied's sister facilities, and eventually, the executives at the corporate level. Long story short, in 1972, Charles J. Latino was the founder and Director of the Allied Chemical Corporate Research & Development (R&D) Reliability Engineering Department, the first of its kind in the U.S. for a major chemical conglomerate. Their research focused on Equipment, Process and Human Reliability (at a time when Human Reliability wasn't a consideration for most companies). At that time, this young boy with a vision of what could be, became known as the Father of Manufacturing Reliability and pioneered the field of Reliability Engineering for future generations to continually improve.

As an FYI, Allied Chemical eventually acquired Honeywell and took their name. Honeywell was a much more recognizable name in the industry.

In 1985, Charles retired from Allied and purchased his department from the corporation. At that point, he could apply his Reliability principles to any industry he wanted to ... and he did, around the world for 30 more years. He did not have the luxury of STEM programs back then, but he certainly proved he was an advocate for them. I appreciate your patience with my wanting to give you some context, but I'm finally getting to my point. As you can see, I literally grew up in a Reliability household. Reliability was not just a job for Charles, it was a way of life. Reliability focuses on identifying high risks and making sure they don't happen. It didn't matter whether Charles was at work, or at home, he was a natural proactive thinker. We even had Reliability-related features in our home. I remember Charles wanted a fire escape in his new dream house in the 60s, just like the steel ones they had on his building when he grew up in New York. When he went to the building code people in the small city of Hopewell, no one knew what he was talking about. Such codes did not exist ... but they soon would. So, I grew up being conditioned that proaction was the norm, and not the exception.

## 1.2 HOW THIS EXPERIENCE HELPED ME RAISE MY DAUGHTER

I worked side-by-side with my father at Reliability Center, Inc. (RCI) for 22 more years until his passing in 2007. We were a family-owned company, so I had the additional pleasure of working with most of my siblings every day. Reliability was in our DNA and it was/is a way of life for us all. We are now passing this gift along to our children.

Because of how I was raised, I wanted to pay it forward and ensure that my daughter received the same opportunities that I did, in terms of proaction being a way of life. I had been trying to work with our local school systems to help kids learn about the technologies that we were imparting on Fortune 500 companies. However, I found most of the school systems to be very rigid in their curriculum and focusing on complying with all their rules, and not willing to try new things the kids would actually use when they got into the workforce. This was at a time when STEM was not as popular as today, where there are progressive school systems who are more receptive to such new technologies.

So, I decided to do my own pilot project with my daughter at the time, who was a fifth grader in elementary school. I wanted to conduct a test about the usability of a new RCA software project our company had developed called PROACT. Understanding she was only in the fifth grade, I wanted to see if I could generate an

RCA success on an issue that would be important to her. She chose to analyse 'bullying', which was very interesting to me, because I was NOT a fifth grader and wouldn't have thought of that … so it was a perfect, relevant topic. I provided her the basics about understanding how to construct cause-and-effect relationships using a logic tree in the software, that we'll describe in depth, later in this chapter.

I also taught her about what was a hypothesis and how to use data/evidence to prove or disprove them. All-in-all I think I spent 30 minutes teaching her the basics about the thought process and the software. Then, she was off to the races on her own and without dad looking over her shoulder. She did Google searches to find articles that would prove or disprove her hypotheses and help her generate more.

I was amazed at her 'work product' and that a bright, young lady could grasp those concepts so fast and apply them to her world. I knew at that time we had a winner and that our PROACT RCA approach and software was applicable to any undesirable outcome, and effective analysts were not limited by their age. It was an epiphany for me.

## 1.3 A FIFTH GRADER'S RCA ON BULLYING

I will now walk you through that actual analysis on bullying. Keep in mind this is the work product of a fifth grader (in 2013), primarily using their internet search results to word their hypotheses.

The 'Event' in this case is the undesirable outcome. Some articles found cited a statistic that stated, 'Bullying makes their victims less self-confident and 9x more likely to commit suicide'. A very sad statement indeed and exactly why issues like this should be analysed in-depth.

The 'Mode' in this case is the act of Bullying. Just remember that from level-to-level, these represent cause-and-effect relationships.

In Figure 1.1, it is determined through the online research results that (at that time) there were three primary forms of bullying:

1. Cyber bullying (4%)
2. Physical bullying (15%)
3. Verbal bullying (48%)

For the sake of this example, we will only follow the most impactful hypothesis, which is verbal bullying. As stated earlier, the fifth grader in this case would just do Google Searches to find papers related to subjects she was looking for. In Figure 1.2, this is what we call a verification log. This just stores the information related to how they proved or disproved that hypothesis. In this case, we can see an internet search was conducted on January 20, and an article was found, which is attached to the record that supports the percentages above.

Moving on down the logic tree we ask, 'How could verbal bullying occur?'. In order for bullying to occur, there has to be three factors as shown in Figure 1.3:

1. A bully
2. Someone being bullied
3. An environment in which to be bullied

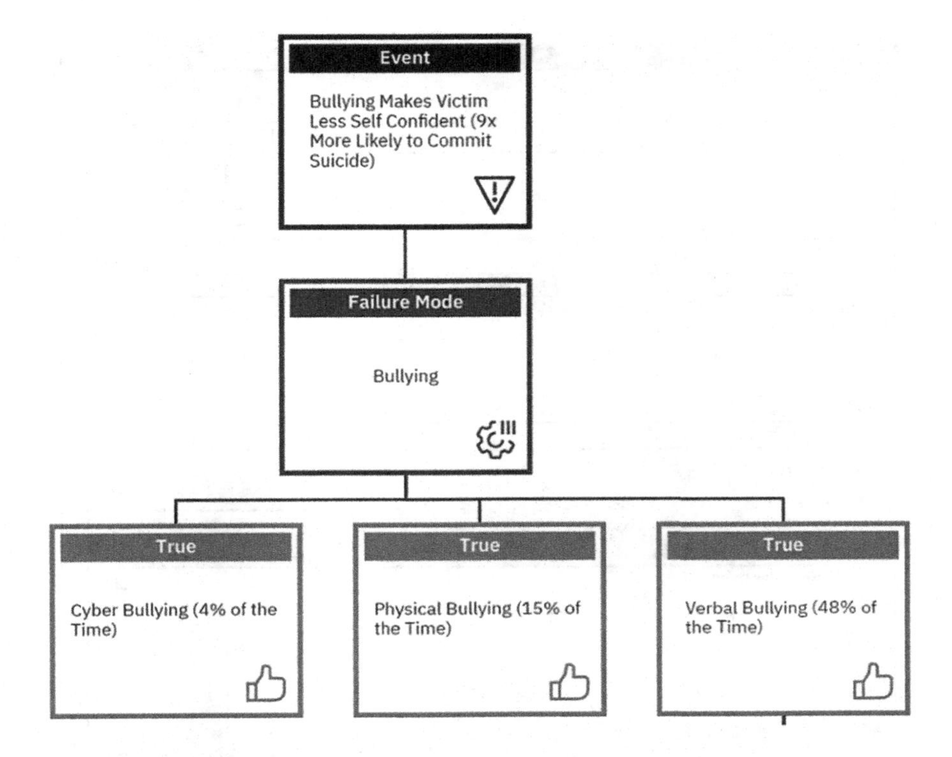


**FIGURE 1.1** Bullying logic tree (1) [Logic tree showing the top level].

So, these factors will be further explored.

Let's start with the bully themselves. Because we are at the point of a person making choices, we will switch our questioning to 'Why' instead of 'How Could'. So, in this case, 'Why do bullies bully others?'. The research concludes the bully:

1. Is insecure
2. Needs to be in control
3. Is in need of attention
4. Is rewarded materially or with prestige
5. Can't control their emotions
6. Lacks the capacity for empathy

Figure 1.4 shows how our logic tree continues to explore deeper roots.

At this point, we will delve deeper into these hypotheses. In the analyst's research, they found very similar factors as to why bullies tend to have these traits, most all rooted in their home life.

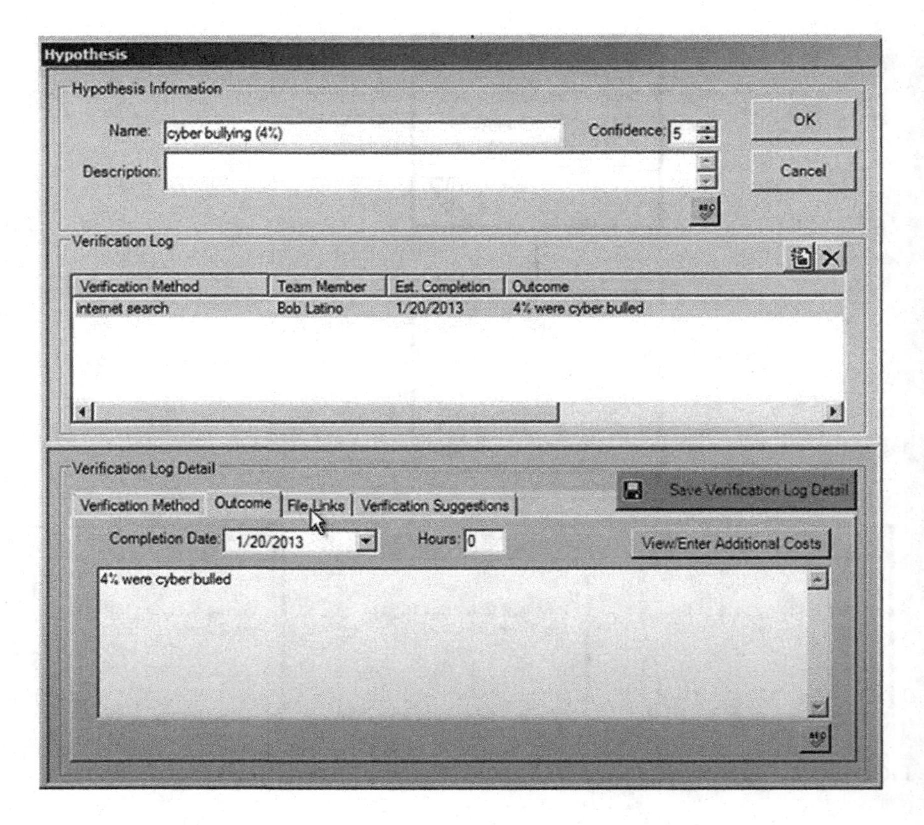


**FIGURE 1.2** Bullying logic tree (2) [Logic tree showing how proven hypotheses are shown].

Here is a listing of such issues in Figure 1.5, from their home lives:

1. Little warmth or affection
2. Trouble sharing feelings
3. Inconsistent discipline
4. Little monitoring of children by their parent(s)
5. Punitive and rigid discipline by parent(s)
6. Usually had physical punishment

So, all of these were consistent for the rest of the bully's traits listed in Figure 1.4.

Let's continue on and now explore why the person being bullied allows that to happen.Not surprising, we find the following from Figure 1.6, such children:

1. are too scared to respond/fight back
2. have low self-esteem
3. have neurodevelopmental conditions

Bullies tend to prey on such children they perceive as vulnerable and a low risk of threat to them personally.

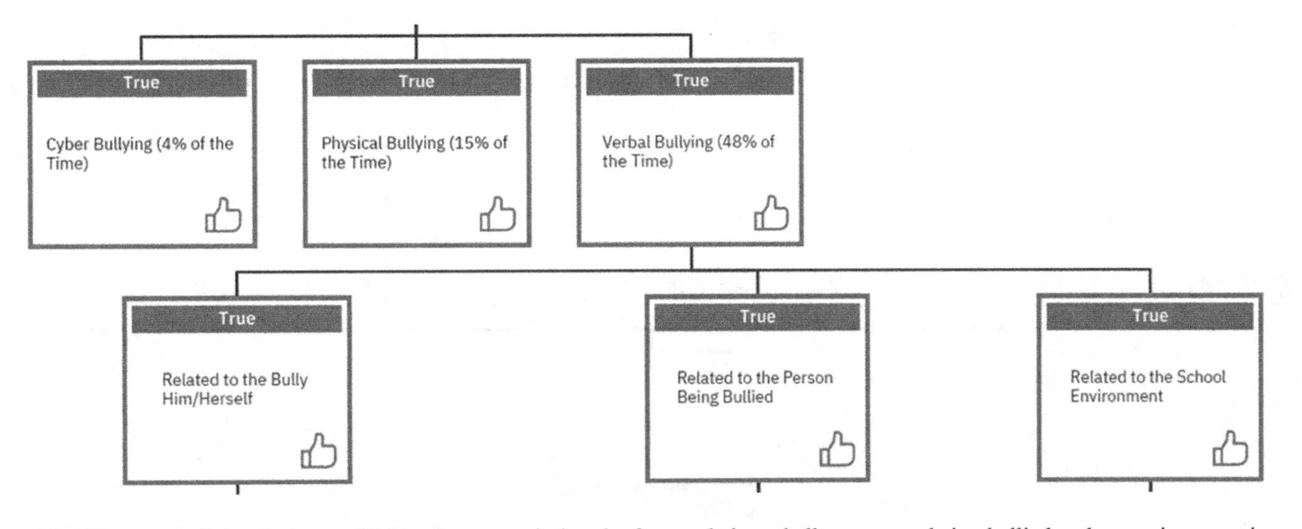


**FIGURE 1.3** Bullying logic tree (3) [Logic tree exploring the factors; being a bully, someone being bullied and an environment in which to be bullied].

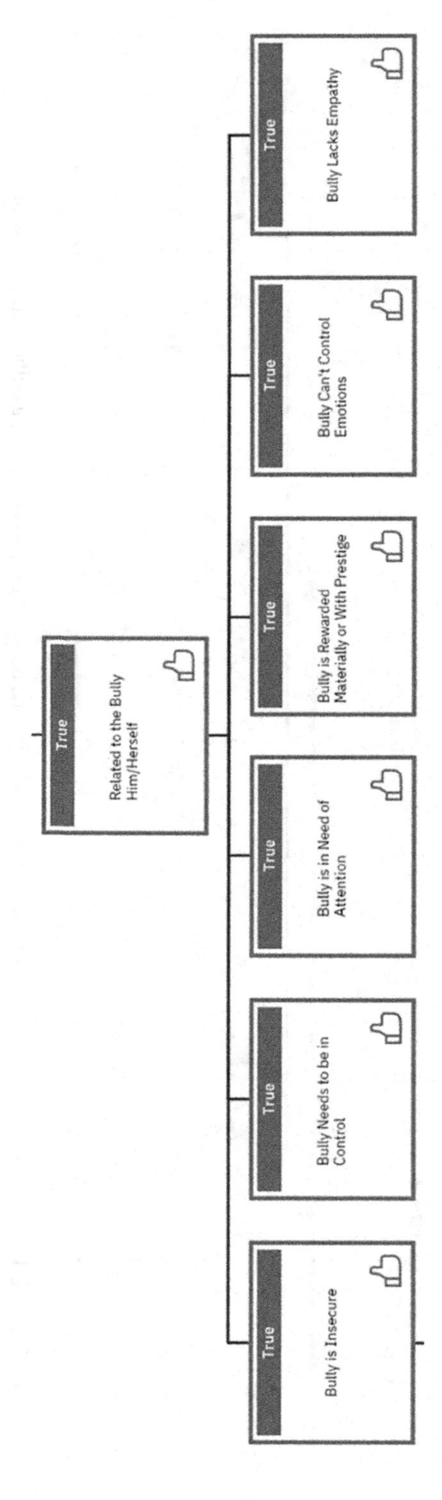


**FIGURE 1.4** Bullying logic tree (4) [Logic tree exploring 'Why do Bullies bully others?'].

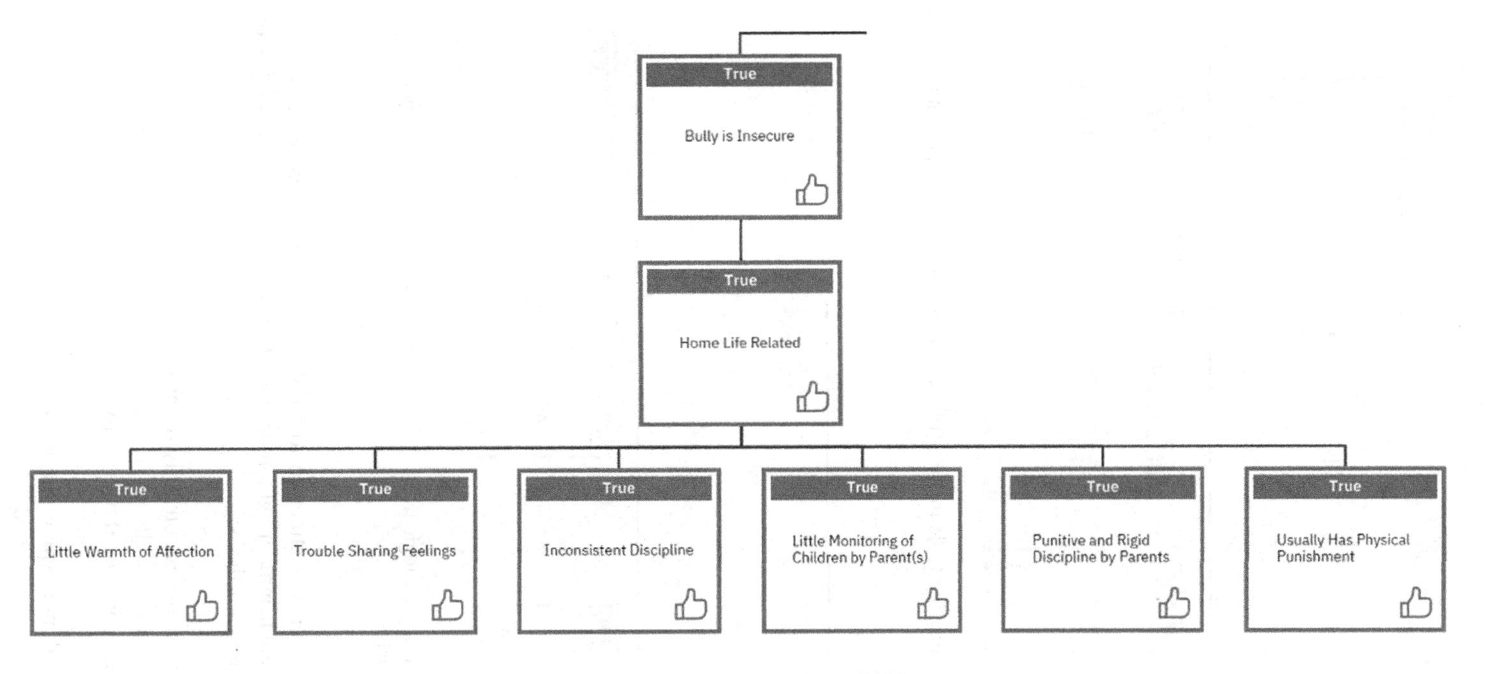


**FIGURE 1.5** Bullying logic tree (5) [Logic tree exploring a bully's environment (home life)].

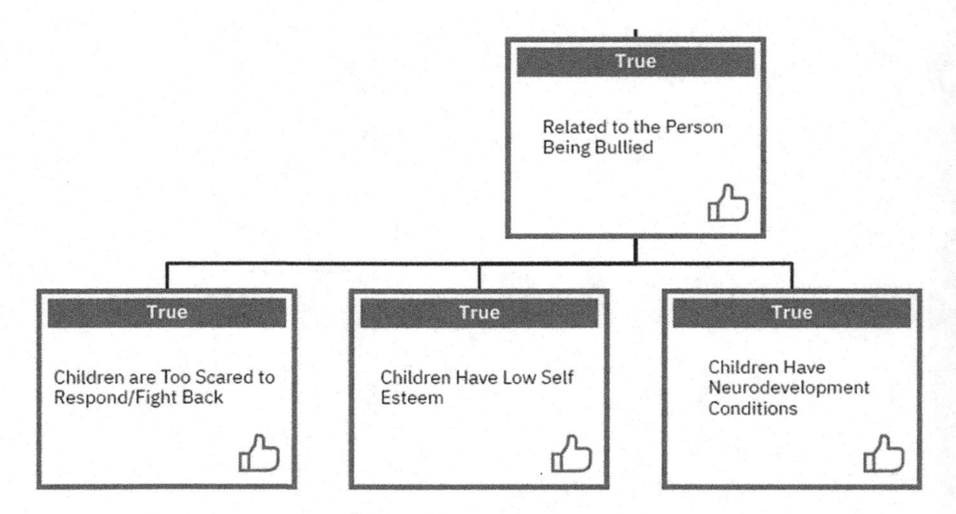


**FIGURE 1.6** Bullying logic tree (6) [Logic tree exploring why a person allows themselves to be bullied].

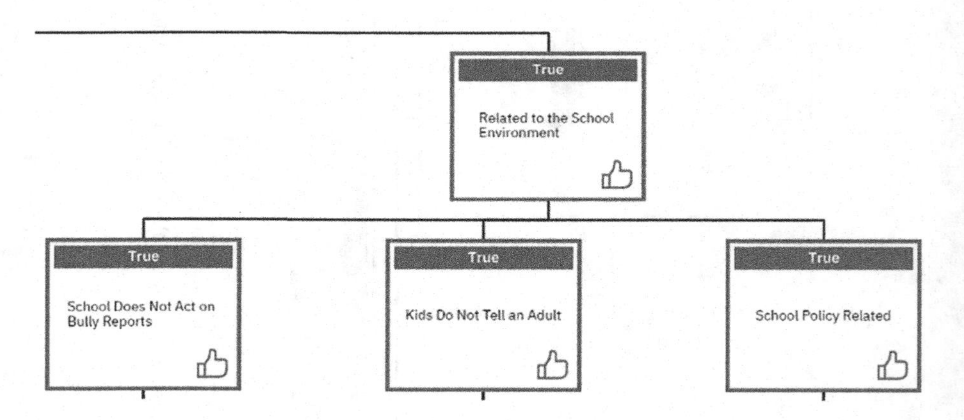


**FIGURE 1.7** Bullying logic tree (7) [Logic tree exploring the environment, which permits bullying to occur].

Now let's divert our attention to the third element needed for bullying to occur, and that is an environment, which permits it. In Figure 1.7, we see a few hypotheses for this, they are as follows:

1. School does not act on bully reports
2. Kids do not tell adults
3. School policies could be a factor

The analysis will not explore these possibilities.

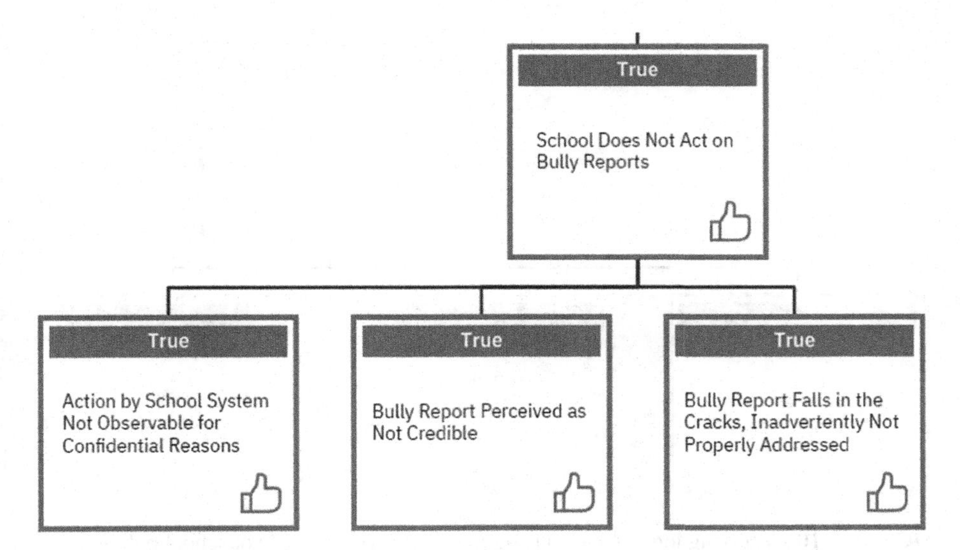


**FIGURE 1.8** Bullying logic tree (8) [Logic tree exploring why the school will not act on bully reports].

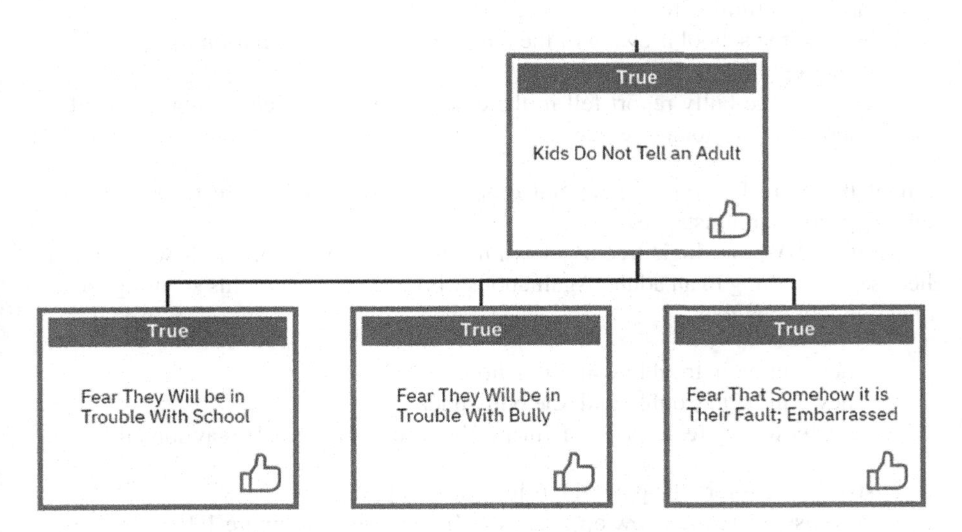


**FIGURE 1.9** Bullying logic tree (9) [Logic tree exploring why kids who are bullied don't report it to an adult].

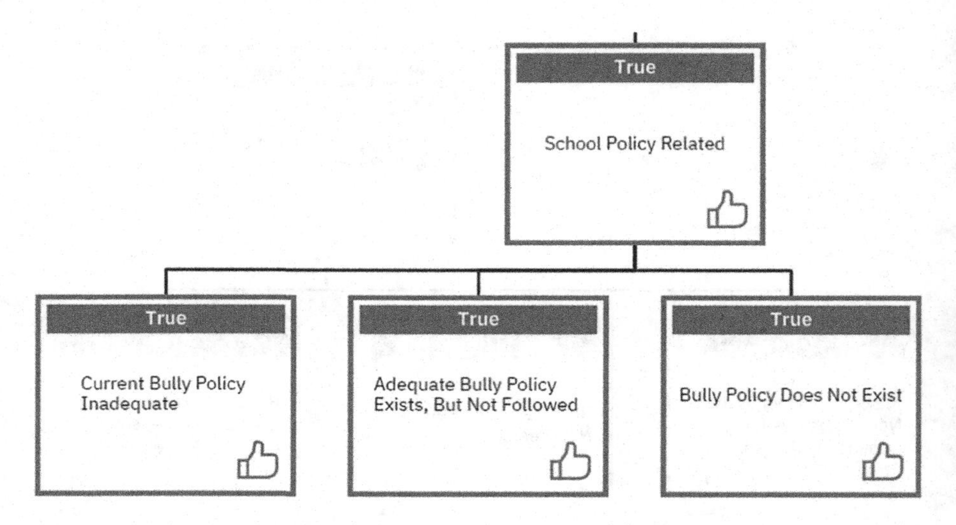


**FIGURE 1.10** Bullying logic tree (10) [Logic tree exploring the role of school policy].

Why could the school not act on bully reports as shown in Figure 1.8?

1. Perhaps the school has acted on them, but because of confidentialities, they are not permitted to say so.
2. Maybe the school is aware of the bully report but have a reason to believe it is not credible.
3. Perhaps the bully report fell in the cracks due to a variety of reasons, and therefore was not addressed.

All of these are legitimate, potential root causes that would have to be validated under specific circumstances.

Moving down our logic tree as shown in Figure 1.9, why don't kids who are bullied, say something to an adult? Again, not surprising answers to this seemingly easy question, but they fear:

1. They will be in trouble with the school.
2. They will be in trouble with the bully and/or.
3. Somehow, they feel it is their fault and they are embarrassed to say something.

Lastly, let's explore the potential role of school policy.

In this case, there are a few options as well, expressed in Figure 1.10:

1. Current bully policy inadequate.
2. Current bully policy is adequate, but not followed.
3. Bully policy does not exist.

So, this concludes a fifth grader's RCA of bullying through her eyes at the time. I'm proud to say it was my fifth grader who completed this analysis. I can also say that the skills she learned from doing this project benefit her every day now because she thinks proactively, using cause-and-effect blocks in her mind, when she is thinking things through. In my book, she's a STEM success story already!

## 1.4 GETTING INTO THE WEEDS ABOUT DEFINING 'WHAT IS A ROOT CAUSE ANALYSIS?'

OK, we started off with applying the methodology on a bullying case, but now let's go through and get into the details that will explain exactly what the RCA approach is and how it can benefit you.

Let's start off with some honesty about the stigma of the term 'RCA' ...it is quite vague, misleading and easily misinterpreted by those who are not immersed in its use. It is a useless and counter-productive term because there is no universally accepted, standard definition. Therefore, any process/tool someone is using to solve a problem is likely to be labelled as 'RCA'. It could be troubleshooting, brainstorming and/or some other more structured problem-solving approaches such as 5-Whys, fishbone diagrams, causal factor trees and/or logic trees.

Ron Butcher, Director of Health and Safety, U.S. Competitive Generation, for AES stated, 'I think the greatest single challenge to an effective causal analysis process, from an organizational perspective, is the focus on the word "root"'. The acronym, unfortunately, does not do the field justice because it connotes there is always a single root cause, and that just does not reflect reality.

If the stigma of 'RCA' is so bad, why use it? One reason is that from a business standpoint, target markets will continue to do their due diligence when selecting qualified providers by doing their web searches on the term 'RCA'. If an RCA provider were to change the analysis name in an effort to create a marketing uniqueness, this means their target market would have to be aware the new term exists (which in reality would be riskier than just accepting the mediocre term of RCA and being in the mix of the search results).

For the purposes of this chapter, I will use my definition of RCA which is:

> 'The establishing of logically complete, evidence-based, tightly coupled chains of factors, from the least acceptable consequences to the deepest significant underlying causes (Latino, R.L., 2011, p. 15).'

The above is a mouthful of seemingly intimidating engineering jargon, but when broken down into practical principles...I promise it's pretty easy. Let's see if I can come through on that promise:

**Logically Complete**: As you will learn in this chapter, this is the difference between asking the question 'How Could?' something happen versus 'Why?' something happened. Prove this to yourself by asking 'How could a crime occur?' versus 'Why did a crime occur?'. The answers tend to be very different. How Could? looks at more possibilities than the narrow and limiting question of Why? (which connotes a singular answer and an opinion).

**Evidence-Based**: This simply means we used hard evidence to prove and disprove our hypotheses instead of letting hearsay serve as a form of validation.

**Tightly Coupled Chains of Factors**: These are fancy words for 'cause-and-effect relationships'. As you will see when we delve into a logic tree, the blocks from level to level, represent cause-and-effect relationships.

**Least Acceptable Consequences**: This depicts our level of tolerance when it comes to what is a 'failure'? At what point will we determine an RCA will be required? When someone is hurt or killed? When we've lost >$1M? When we have been fined by regulators for violations? What are our least acceptable consequences?

**Deepest Significant Underlying Causes**: This is simply at what point will we stop drilling down? When is it good enough to stop? Do we stop when we find the parts that broke? Do we stop by blaming people for bad decisions? Do we stop after understanding the human reasoning for poor decisions, and the sources of flawed information fed to the decision makers?

So, after breaking that complex definition down, it's not all that complicated after all.

### 1.4.1 Introducing the Logic Tree

As mentioned earlier, our failure reconstruction tool of choice for this chapter will be what we have been calling a Logic Tree. Logic Trees. The PROACT® (PROACT is a registered trademark of Reliability Center, Inc.) Logic Tree is representative of a tool specifically designed for use within RCA. The logic tree is an expression of cause-and-effect relationships that queued up in a particular sequence, at a particular time, to cause an undesirable outcome to occur. These cause-and-effect relationships are validated with hard evidence as opposed to hearsay. The evidence leads the analysis, not the loudest expert in the room.

A logic tree starts off with a description of the facts associated with an event. These facts will comprise what is called the Top Box (the Event and the Modes). Modes are the manifestations of the failure, and the Event is 'the least acceptable consequences' that triggered the need for an RCA. While we may know what the Modes are, we do not know how they were permitted to occur. So, we proceed with the questioning of 'How Could' the Mode have occurred?

## 1.5 FUNDAMENTALS OF A LOGIC TREE

**Cause-and-Effect Logic**: From level to level, it represents a cause-and-effect relationship. This does not have to be a linear relationship, as there may be multiple causes that have to occur at the same time, in order to create that effect. We just need to know that we are simply creating a graphical expression of logic to reflect the facts that occurred, to cause an undesirable outcome.

**Event**: The reason you care! What brought this incident to your attention? Many believe that we do RCA on incidents themselves, but I believe we do RCA on their consequences. Think about it at your place, there is usually a business level reason we do RCA…injury/fatality, certain $ production loss exceeded, certain $ maintenance cost exceeded, regulatory violation and the like (often called triggers). Sound familiar? These are the known FACTS.

**Failure Mode**: These are the typical things we normally start an RCA with, like Pump Failure, Injury, Loss of Production, Environmental Excursion, etc. These led to the Event. These too are FACTS.

**Hypotheses**: Just like in high school, these are 'educated guesses. These are potential causes to the preceding nodes. The initial question after the Failure Mode is 'How Could the preceding node have occurred?'

**Verifications**: These are the ways in which we proved, with sound evidence, that Hypotheses were True or Not True. Fun fact … hearsay is NOT a valid verification technique.

**Physical Root Causes**: These are where the physics of failure root out. These are observable, tangible things we can see. These are usually the immediate consequences of decision errors.

**Human Root Causes**: THIS IS NOT 'THE WHO DUNNIT'! This is the act of decision-making. These are usually errors of omission and/or commission. We did something we were not supposed to, or we were supposed to do something, and didn't. The key here is to NOT BLAME and take the opportunity to understand human reasoning.

**Latent/Systemic Root Causes**: These are the organizational systems, cultural norms and socio technical factors that influence and contribute to our decision-making. Unfortunately, our 'systems' are far from perfect and are always a work-in-progress. They include, but are not limited to our policies, procedures, training systems, purchasing systems, HR systems, compliance systems and the like.

**Contributing Factors**: Identify items which did not directly lead to the failure but created vulnerabilities allowing the failure to occur. These are usually conditions that we don't have control over, but we can often compensate for them (if we are aware of them). For instance, some failures may only occur when it's freezing outside. This is a condition that we can't change, but we can compensate for them in order to mitigate their potential consequences.

I would like to note that in our bullying example, we did not label any actual root causes (physical, human or latent), because the logic was followed generically and not related to a specific case. The bullying case is a great example of using RCA to construct troubleshooting flow diagrams for use on potential failures in the future.

In our last section, we will apply the same RCA approach, to a specific failure in a manufacturing plant. In this case, we will label-specific root causes.

## 1.6 BRINGING PROACT RCA THINKING INTO A MANUFACTURING PLANT

Now that we've demonstrated how to apply the PROACT RCA thought process to a softer, administrative type of undesirable outcome like bullying, let's now bring it into an actual manufacturing plant and apply it to something mechanical and a bit more complex.

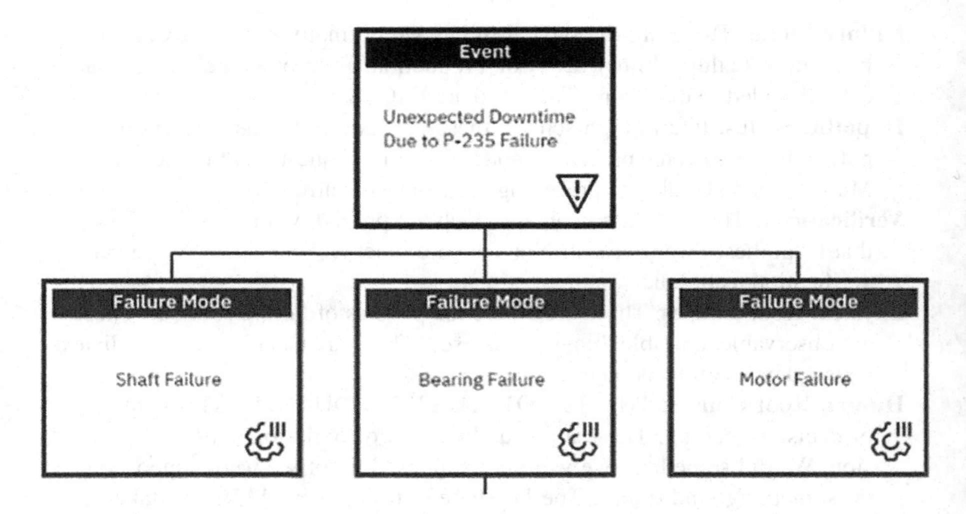


**FIGURE 1.11** Event + Modes = Top Box [All must be FACTS] [Top of the logic tree showing the event and modes which must all be facts].

In Figure 1.11, as mentioned earlier, the EVENT is the reason we care enough to commission an RCA. In our example, the event is 'Unexpected Downtime Due to Pump-235 Failure'. Now, the MODES are going to be how we have experienced such failures in the recent past. Most of our CMMSs can produce these high-level modes. In this case, such downtime due to this pump failure has been attributed to failed shafts, bearings and motors. Our data can also tell us the annual cost of each of these modes (hopefully downtime $ + labor $ + materials $). In our case, we know that bearing failures on this pump represent the most annual costs. To make a quick business case for your failure, try out this free chronic failure calculator (CFC).

The Mode level is what we consider our FACT LINE to start. If we start with facts, and provide our hypotheses with sound validations, we will end with facts. Keep in mind we are travelling down the path of the physics of failure, so we will continually ask the same question, 'How Could'.

As you use a logic tree to explore the physics of failure, imagine you have the luxury of a video recorder in your head, and you are watching the event as it's played in reverse. In our case, 'be the bearing'. Ask yourself, 'How could I have just failed?'. Move back in short increments of time. It takes some getting used to this type of thinking, but that is the beauty of the logic tree, it guides us without any biases. This tool, when used properly, should be non-personal and non-threatening. We are interested in valid hypotheses, possibilities … that's it. Then, we will use evidence to demonstrate which hypotheses were true and not true. We will only continue drilling down on the one that was not true.

In our case, based on the SME (subject matter experts) on our team, we conclude there are only four (4) ways in which a component can fail: Erosion, Corrosion, Fatigue and Overload. So, we list them as shown in Figure 1.12.

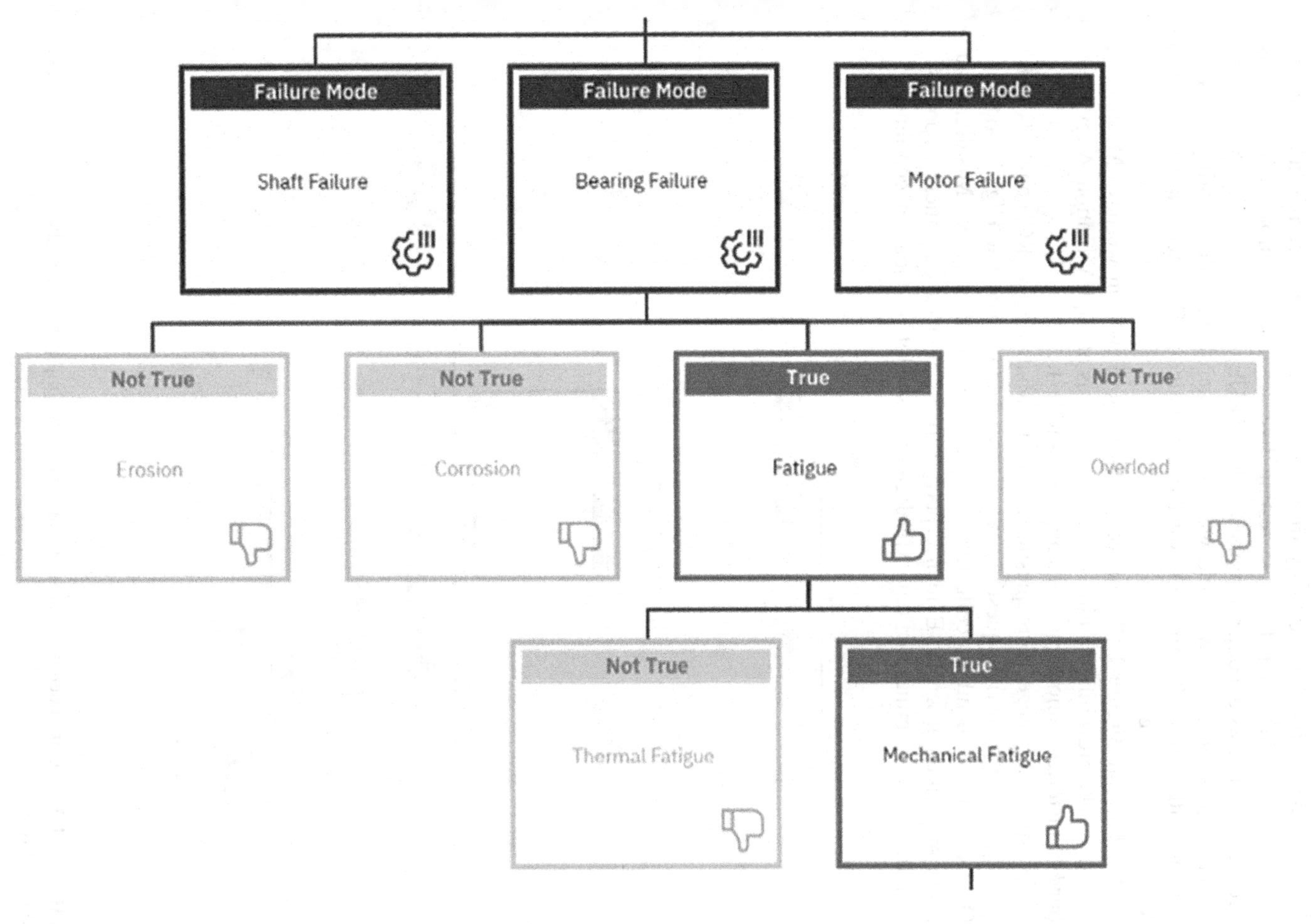


**FIGURE 1.12** Hypothesizing and validation [Logic tree showing how a hypothesis appears after it is verified].

In our example, we have our on-staff metallurgist visually inspect the failed bearing. They determine with certainty from a visual review, the bearing failed due to Fatigue. No additional exhaustive testing like scanning electron microscopy is needed. This makes the other hypotheses NOT TRUE.

Same questioning, 'How could we have fatigue of the failed bearing?'. SMEs indicate either from Thermal or Mechanical Fatigue. The metallurgist confirms Mechanical Fatigue.

As shown in Figure 1.13, our team now asks, 'How can we have Mechanical Fatigue?' The prevailing opinion is a sole hypothesis of High Vibration. A review of our PM histories demonstrates this hypothesis to be true.

Questions only beget more questions, as that's what effective RCA analysts do for a living; they ask the right questions. So, 'How could we have had High Vibration?'. Our RCA team members collectively come up with: Resonance, Misalignment, Imbalance and Looseness. Evidence pooled together to validate these hypotheses and the team determines that only Misalignment is valid. The journey continues!

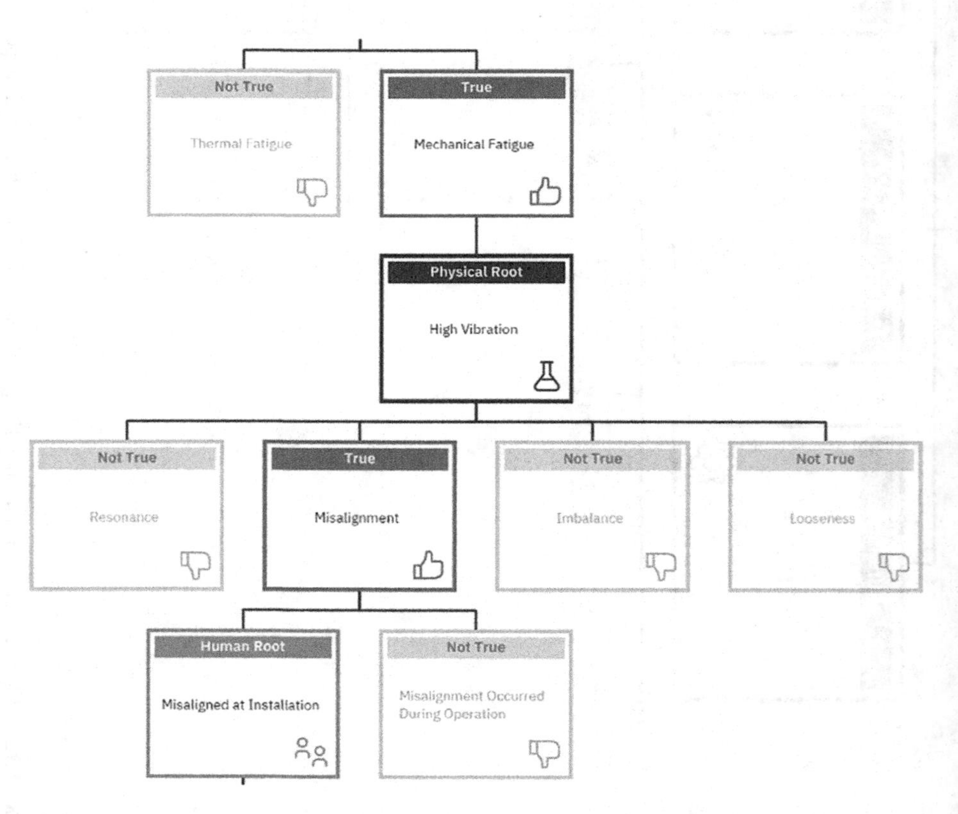


**FIGURE 1.13** Continued hypothesizing and root labelling [Logic tree showing the different types of roots].

How could we have ended up with misalignment? This is where we are now crossing over from the physics of failure, to the human and systems side of failure (or the social sciences). Either someone misaligned the pump from initial installation or repair, OR it was aligned correctly and then became misaligned in operation. Vibration histories demonstrate that since the last installation, this pump has chronically had vibration issues. Notice here that we switched the label on the high vibration node from a hypothesis to a physical root. This is because this is the first visible consequence after the triggering decision.

Notice that after the decision point, everything is triggered on its own, as cause-and-effect linkages go into play. If there are no human interventions to break the error chain, then it will play out and contribute to the undesirable outcome (Event).

We are at a pivotal point in our logic tree at this time. Why? Because we have uncovered a decision point. The mechanic in our case chose to align the way they did, on that day. A 'decision' point is our queue to identify a human root, and to switch our questioning to 'Why' instead of 'How Could'. We are not interested in learning about the infinite reasons the human 'could have' made a decision, we are interested in 'why' they did. This is also the point in the logic tree, which switches from deductive reasoning to inductive.

So, let's drill down further and see if we can figure out what was going on in the mechanic's mind that day!

So, in Figure 1.14, after interviewing our mechanic (using human performance interviewing techniques), we uncover many things that we did not know.

1. The senior mechanic retired, and part of his duties were conveyed to the junior mechanic that did not have the training and certifications to align properly. They did the best they could with what hand they were dealt.
2. The alignment tools provided were less than adequate (LTA), representing old technologies.
3. There was LTA management oversight in the sense there was an unqualified mechanic aligning critical equipment. Where were the checks and balances to prevent that?
4. Finally, we find out that the trade-off pressures between operations and maintenance/reliability influenced the decision. There was significant pressure to get the process running again ASAP. When that typically happens, the natural human tendency is to take short cuts. This usually comes in the form of skipping steps in a sequence of tasks we must do.

Chances are, all of these 'Systemic Roots' have contributed to other failures as well individually or in combinations. This is because most systems are put in place for a multitude of people to use, under a variety of conditions. This particular combination of system flaws converged on this day, to influence the well-intended mechanic's decision that day.

Free suggestion, if it was happening to this mechanic, we should check the skill level of others who may be victims of their own systems as well. This is when we

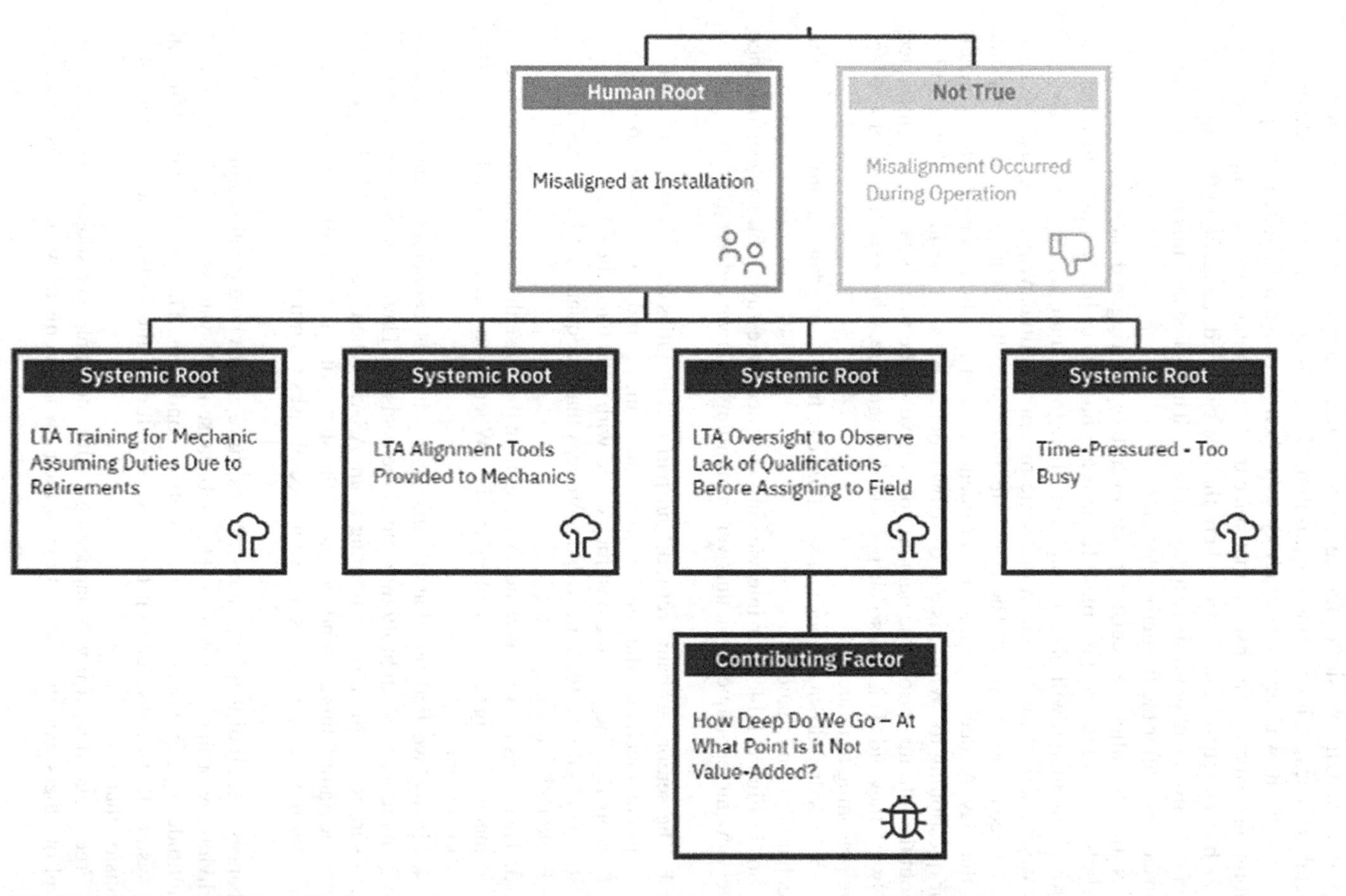


**FIGURE 1.14** Continued root labelling [Logic tree showing more info about the mechanic and situation].

determine if the recommendation/correction action is isolated to a single case or more universal.

Lastly, I put a lingering contributing factor in our logic tree to make a point. When do we stop digging?

My personal rule-of-thumb to answer this question is, 'When the solution is obvious'! I tend to not drive down to see who set up the flawed system because I don't care. It is not value-added because what benefit do I get from finding that out (especially if they are not there anymore). Also, if drilling deeper gets into issues outside our fences, is it of value? In most cases, we will not have control of things like changing regulations. However, in many cases if we see an OEM design flaw, we may opt to have our engineering department pursue that path with the OEM. But as the analyst in the field, I can hand that part off and move on to my next RCA.

Is everything covered in this chapter about 'RCA' … absolutely not. I will post some links at the end of this blog where you can learn more about holistic RCA, but I wanted to get you started with the basics.

Figures 1.15A, 1.15B and 1.15C are presented as simple job aids to help those starting out on the field as RCA analysts, learn the fundamentals of constructing a logic tree. I hope you find this job aid of value.

Remember, 'We NEVER seem to have the time and budget to do things right, but we ALWAYS seem to have the time and budget to do things again!'. Let's do RCA right the first time so we don't have to analyse the same Event again😊!

Additional RCA-related Information:

1. ***Root Cause Analysis: Improving Performance for Bottom-Line Results (5th Ed). Latino, Latino and Latino*** **(332pages)** https://www.amazon.com/Root-Cause-Analysis-Performance-Bottom-Line-ebook/dp/B07TLG2HBM/ref=sr_1_1?dchild=1&keywords=Root+Cause+Analysis%3A+Improving+Performance+for+Bottom-Line+Results%2C+Fifth+Edition&qid=1594216642&s=books&sr=1-1
2. ***The PROACT Root Cause Analysis Quick Reference Guide (Latino, Latino & Latino). A summarized Focus Book (only 82 pgs)*** https://www.routledge.com/The-PROACT-Root-Cause-Analysis-Quick-Reference-Guide/Latino-Latino-Latino/p/book/9780367517380
3. ***Reliability Center, Inc.*** **–** www.Reliability.com
4. ***EasyRCA Software*** – www.easyrca.com
5. ***Reliability Learning Center*** – https://learning.reliability.com/collections
6. ***Bob Latino Blog Posts*** – https://www.linkedin.com/in/robert-bob-latino-3411097/detail/recent-activity/posts/
7. ***Video Case Study*:** https://www.youtube.com/watch?v=1vnsUxofIUg&t=2s

As always, I am available via LinkedIn (see above) for help with any RCA-related questions you may have.

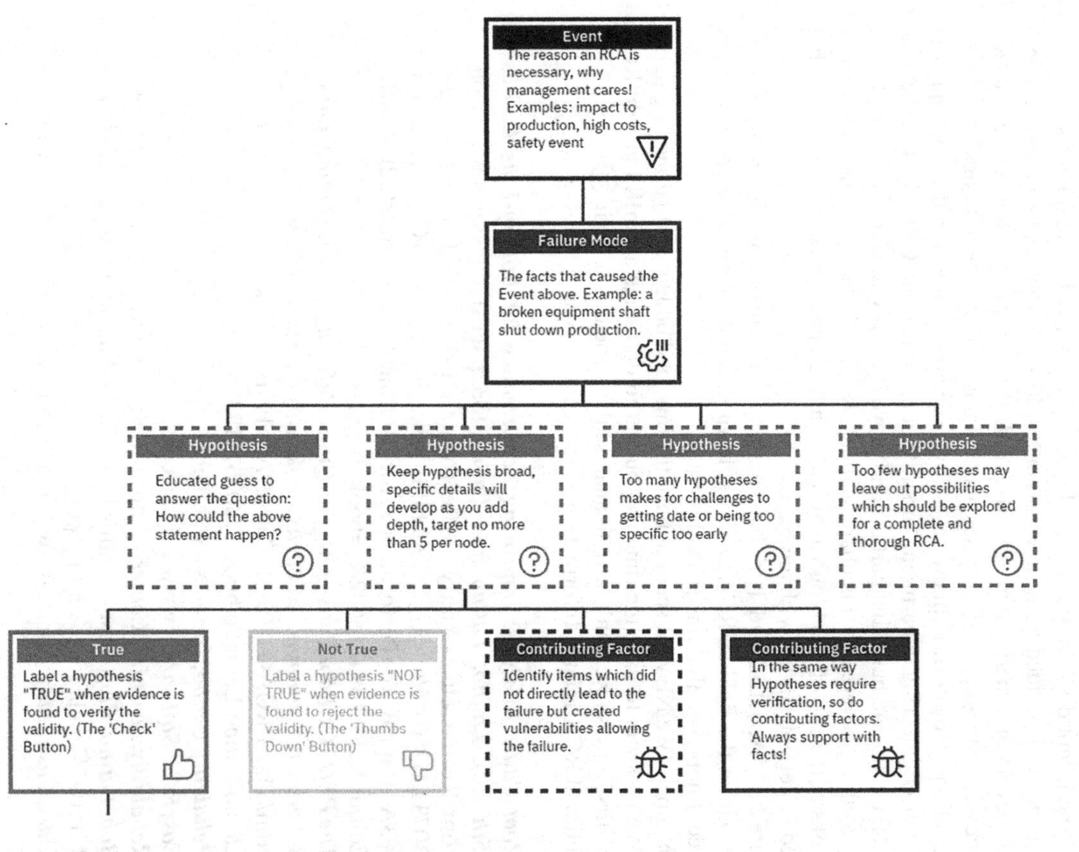


**FIGURE 1.15A** PROACT logic tree reference guide (Part I) [Logic tree – job aid part 1].

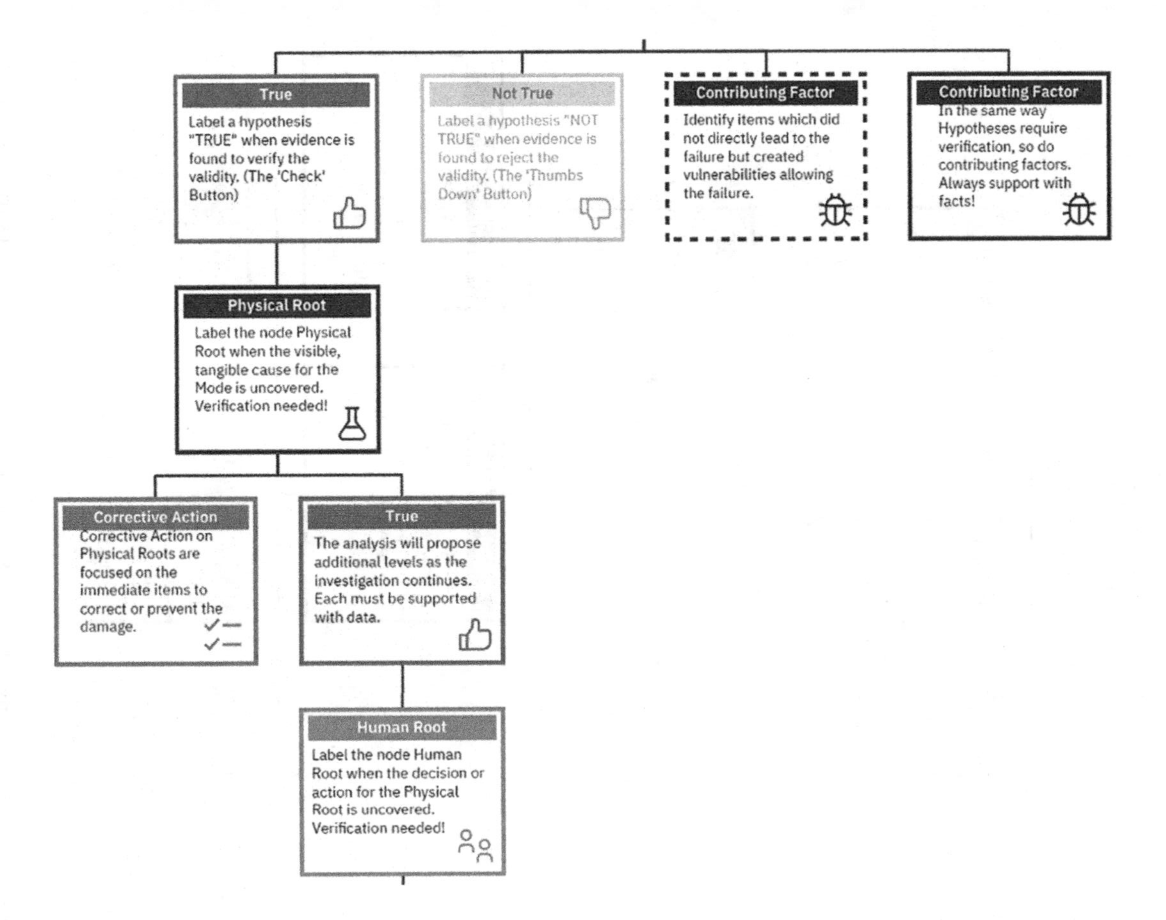


**FIGURE 1.15B** (Continued) PROACT logic tree reference guide (Part II) [Logic tree – job aid part 2].

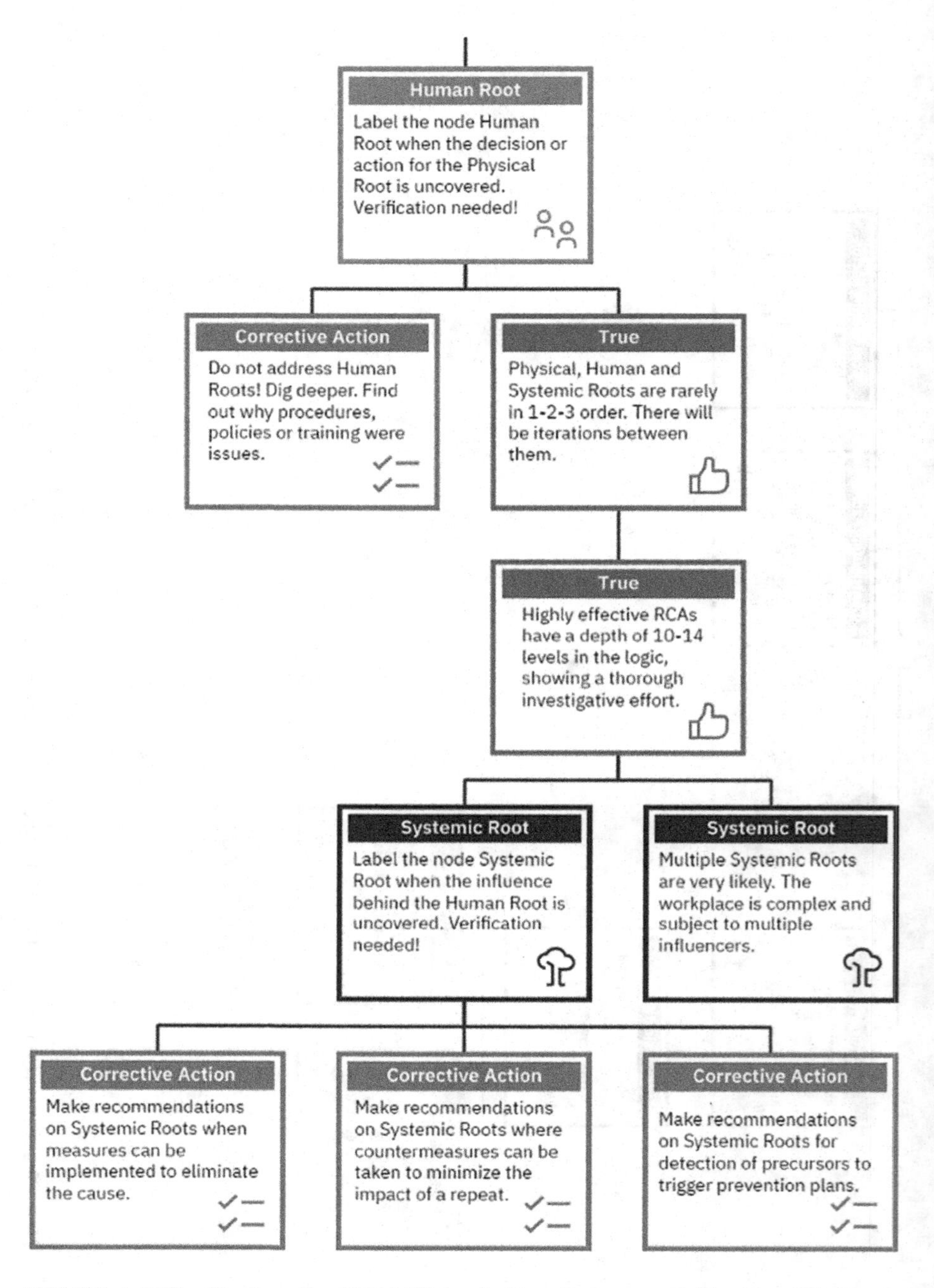


**FIGURE 1.15C** (Continued) PROACT Logic Tree Reference Guide (Part III) [Logic tree – job aid part 3].

## ABOUT THE AUTHOR

**Robert J. (Bob) Latino** is currently a Principal with Prelical Solutions, LLC. Bob was CEO of his family's 49-year-old Reliability consulting firm, Reliability Center, Inc. (RCI) until it was acquired in 2019. Bob has been facilitating root cause analyses (RCAs) analyses with his clientele around the world for over 35+ years. He is author or co-author of seven (7) books related to RCA, Reliability and/or Human Error Reduction. He is an internationally recognized author, trainer, software designer, lecturer and practitioner of best practices in holistic RCA systems (encompassing equipment, process and human reliability).

# 2 Navigating the Path Not Usually Followed

*Emma Holloway*
Colorado State University, United States of America

**CONTENTS**



Growing up, I thought everyone had the same love of science as I did; when watching TV shows or reading my favorite books, I was enthralled by the experiments kids my age were allowed to engage in. Going into more involved science, I learned this was not the case for most my age. Discussions surrounding math, science, or technology were not at the forefront of dialogue. Instead, these discussions seemed to isolate me. My first experience with this segregation was in kindergarten when I was placed into a group called the "LEAP kids."

## 2.1 LEAPING FORWARD

Teachers or parents could sponsor or elect their children to testing that could place them into a special program known as LEAP. LEAP, which is an acronym for the values of the program, allowed students to leave class during set times throughout the week to engage in "diverse" learning. At first, I was the only girl in the program. As I progressed through elementary more students joined, and I had other female counterparts. Through my time in LEAP, I never once felt like I did not belong. I was able to solve all the problems, aced my projects, and nailed the various presentations. The times when I felt like an outsider occurred when I was placed in my regular class. Other students seemed mad that I was given time to "goof around" when I should have been doing multiplication tables or class reading. While I wish I could say this – "seeming" – jealousy subsided as I grew older, it did not.

Middle school brought forth a new set of challenges as, instead of a more casual title, the LEAP kids transitioned into the GT kids. GT stood for gifted and talented, which as you can suspect puts a target on your back in the everchanging junior high halls. Never once was I seriously bullied, rather I was put into separate classes according to my newfound title and advanced through the math. Since you had to test into this program in elementary school, many kids did not have the chance to enroll in the special classes. This changed when the transition from middle school to high school occurred.

DOI: 10.1201/9781003336495-3

## 2.2 STAYING AHEAD OF THE CURVE

At north Texas high schools, academia is huge. Everyone knows your GPA and class rank at the beginning of the first day of freshman year. To get ahead of the curve, students sign up for as many AP (Advanced Placement) classes as possible as soon as possible. There are no prerequisites to AP classes, which invite a variety of students to sign up for them. I understood these types of classes to have better teachers and dive deeper into the content which I adored; however, I quickly understood this was not the mentality for many other students.

A common theme evident in my AP classes was a level of arrogance for the students. Even the ones who would fail nearly every assignment would brag to whoever would listen about their status as an "advanced placement" kid. This idea, I realized, was held by many of those in each of the groups I was a part of during my K-12 years. There were those who believed because they could complete a difficult problem or read beyond their reading level, they were better than everyone else.

Arrogance often leads to rage. Every experience where I have felt discriminated upon as a woman in STEM has been fueled with the idea that someone *should* be better than me; that because I am a girl, conventionally pretty, and outgoing I shouldn't be able to outdo them. This understanding became most apparent to me in my upper-level classes, specifically in high school. I took advanced placement chemistry in my sophomore year with many other sophomores. During the class, we were placed into random lab groups to work with; I was put into a group of girls who were known to be intense regarding their studies. As we went through the lab, there was a subsection that the others were confused about, but I was not. Being very confident in chemistry and having one of the highest grades in the class, I explained the subsection through. Not one of my group members listened. Instead, we waited – around 25 minutes – until our teacher had the time to explain it. When she finally came around, she said the exact reasoning as me. I wish I could write that this arrogant position that some students hold got better when I arrived at college, but unfortunately it didn't.

## 2.3 FOLLOWING MY PASSION

I am not sure of a university where discrimination doesn't occur. Colorado State University works hard at pushing a narrative of inclusion and diversity, but bias still exists. There are going to be people along the academic route that believe you should not be there – those who ignore your presence when discussing a topic, those who do not believe you are smart because you don't "fit the profile," and so many more. This, however, is never a reason to decommit from your passion.

Being in STEM has allowed me to have the most beautiful and intelligent friends who value me for my personality, not just my brain. We have lengthy conversations about space travel and chemical reactions we find interesting but can pair these immediately with constant laughter and joy. There will be those who believe you can't or shouldn't pursue STEM because of this, that, and another. So, instead of taking their words as the truth, move past it, and explore what makes you passionate. Passion is the pathway to happy success.

## ABOUT THE AUTHOR

**Emma Holloway** is a student at Colorado State University studying Biomedical and Chemical & Biological Engineering with a minor in Statistics. She is a part of Biomedical Engineering Society, Women in Engineering, and Alpha Sigma Kappa – Women in Technical Studies. In her free time, she tutors peers in chemistry and loves to go on hikes.

# *Section II*

## *Navigating the World of STEM*

# 3 Power to Make a Difference

*Heather Eason*
Select Power Systems, LLC, United States of America

## CONTENTS



So, you think you can't do whatever it is you put your mind to, and that your dream is unattainable? Why? Let me tell you my story and show you it can be done.

Allow me to introduce myself. My name is Heather Eason, and I'm a business owner, an engineer, a mom, a daughter, a sister, and a wife, so I really have done it all. But it didn't come without effort and hard work. Here's a story to better illustrate the events that molded me and influenced my pathway.

I wasn't the typical girl. Against my mom's wishes, I took both Algebra and Geometry at the same time instead of Home Economics. I was in the marching, concert, and jazz bands. My extracurricular activities included being on the chess team. Yes, I was a nerd.

While I was good at math and science, I don't think those two things are what would later define my career path. My Dad was cheap, frugal, and penny-pinching. He didn't believe in paying anyone to do or fix anything that he could. He believed that he, my brother, and I could figure out anything. And we did.

Need to change out the transmission on a car, no worries if you don't have a lift. Arrive home from school with your Dad in the front yard holding shovels, sound weird? Not to me. We dug a hole, pushed the car over it, crawled under, dropped the transmission, pushed the car out of the way, moved the old transmission out, put the new transmission in, pushed the car back over the hole, crawled under, installed it, and drove the car to the parking pad before my Mom got home from work. Needless to say, she was furious that we put a hole in the front yard, but we wanted to use the slope of the yard to our benefit and that was the best spot. We planted a dogwood tree and azaleas there to make her happy.

For a few years, my parents owned a go kart track, my friend and I were their pit crew. I bled brakes, overhauled motors, and changed tires. My friend's grandfather

DOI: 10.1201/9781003336495-5

was worried we would amount to nothing, so he even taught us how to weld, we welded the bumper sections around the track.

We fixed the sewing machine, vacuum cleaner, AC, TV, dryer, and unsuccessfully fixed the hair dryer. I'll never forget it shooting flames at my Mom's head leaving a section scorched. That was the last time we fixed a hair dryer and purchased new ones going forward.

My Dad believed that being a girl made no difference. When we needed to roof the split-level house, we practiced on my grandmother's detached garage first, which meant we were working during the heat of the summer. We got it done without any leaks.

It was the end of November and one of the first chilly days, so I was happy to wear one of my favorite sweaters and a pair of jeans. It was a Wednesday. Why do I remember all of this? Because that day changed my life forever. As a 16-year-old girl, I suddenly faced a horrific event that I had no control over, which would test me to what I thought was beyond my limits. As a result, nine months later, I gave birth to my son. I was 17, a high school senior, had a baby, worked part-time, and had no idea what to do.

When I finished my junior year, I had straight A's and I was determined to have a normal senior year, so it didn't occur to me to change my schedule even though I was pregnant. Wow, that was a mistake!

Summer was pretty normal, I worked part-time as a cashier. This gave me money to go to concerts with my friends, go to Myrtle Beach for a week, and have enough gas money to drive anywhere my heart desired, which was usually some rural, country road with cow pastures.

As summer came to an end, school was about to start and my due date was fast approaching. I had gone past the estimated date, but I felt fine, so I continued life as any high schooler would. My water broke around 6:30 in the morning, but I felt okay so I got up, took a shower, went to the store to pick up some last-minute items, and hit up my friend to go to lunch. She worked for her father, a local primary care physician. When I dropped her off after lunch, Dr. Jones, yes that is truly his name for those Indiana Jones fans out there, predicted that I would have the baby that night. He was right and Kevin was born later that day. There were a few complications, so I had to stay in the hospital a little longer than expected. I also had to stay out of school for a while during my recovery.

Can you imagine trying to learn AP Calculus at home, without any instruction, while breast feeding every 2 hours? Yeah, I didn't learn a thing. When I was able to attend the school in person, I was excited. I thought life would return to normal. Not quite. I struggled just to graduate. I failed AP Calculus, barely passed AP English, and the rest of my college prep classes were C's. I was rocked, my confidence shaken and disappointed. I knew I couldn't handle college right away. I needed to regroup and come up with a plan, so I increased my hours at work and focused on trying to be the best mom I could.

## 3.1 BEING A BUSINESS OWNER, ENGINEER, MOM, AND WIFE

So, how does that get me to being a business owner, engineer, mom, and wife? Let's fast forward several years and you will find me married with a house, four kids (three stepchildren added to the mix), a job, pets, and bills. So, I thought, now is the right time to start college. For those of you that know me, you know I don't shy away from challenges, but looking back on it, I should have had my head examined.

I enjoyed chemistry and math and wondered how I could combine those two, so I started taking Chemical Engineering classes at USC Upstate, 30 minutes from my house. I stayed in that major for two years, but I would have had to transfer to USC, in Columbia a little over 2 hours away, to complete that degree. Did I mention that I had a husband, kids in school, a job, and we had a house? Moving so that I could continue my education wasn't an option. I had to find another route to complete my degree, enter UNC Charlotte. It was over an hour away, so I would have to make the horrible commute up I-85 for every class, and they didn't offer Chemical Engineering. Quitting was not part of the equation, so I changed my major to Electrical Engineering and started the long trek to Charlotte each day.

Why Electrical Engineering you ask? My dad was an operator at the Oconee Nuclear Generating Station while I was growing up. I remember him telling me stories and explaining the work he did, the reaction process, and how power was generated. I was transfixed as I listened to him. I knew then that power engineering would be my future because it was part of my past.

I am being honest here, so my college experience was different from most people. It took me six years to graduate. I had to change schools and majors, commute over an hour just to take classes, had a family, and worked full time. I am very proud of my 3.0 at the end of six years, but I had obstacles. I failed classes. Yes, I did. I got an F in a Physics class and a D in Signals, so I had to repeat both classes. This was in college before colleges would replace grades, so my F and D stayed part of my permanent record.

"Wait, so you failed classes and kept going?" Yes, and I'm not going to tell you it was easy. I felt stupid when I compared myself to my classmates, who I thought were all a lot more intelligent than me. But most of them didn't work or have any other responsibilities other than being a full-time student, while I only could study between classes, during breaks at work, after all the kids went to sleep and we had done all the housework. There was more than one time that I called my husband crying telling him that I just wasn't smart enough and didn't know if I could do it or not. Then, the ex- football player became my biggest cheerleader and helped me push through.

I do have some fun stories of that time, whenever my son's school was out, or I couldn't get a sitter, I would take him to class with me. He became almost a mascot of our study group. One time during the class, Dr. Tsu used my son's markers to explain a concept to the class, which Kevin thought was the ultimate in cool.

I wanted to continue to get a master's degree so I took classes my senior year at that level that would count toward both my bachelor's and master's. However, when graduation happened, I realized it was time to "get a real job" and help my husband more with the finances. So, graduate school would have to wait, but we all know that I will get back to things, I had to keep moving forward.

## 3.2 THE IMPORTANCE OF MENTORSHIP

I wasn't the typical college graduate. I had been working and raising a family, so I wasn't a stranger to responsibility and time constraints. When I started my first job, it was easy to stand out against the rest because they were still trying to figure out life and the "real world" isn't nice. As a result, I was able to be recognized by some amazing people who took me under their wings. My first mentor, Paul, told me to never talk about college sports, religion, or politics. That's one that I tell all my interns. I'll stop and say now, being a mentor is one of the greatest things you can do. It's a way to personally give back and share your life experiences and knowledge, with someone who needs it. You become an ear to listen, hand to pat on the back, and foot to kick them out of the nest when it's time to fly.

As I navigated my career, I found several great people to teach me the ropes. It is amazing how willing someone is to share their knowledge when you show interest and take initiative.

It's not all been rainbows and lollipops throughout my career. I have experienced sexism at its worst on several occasions. Unfortunately, here are several real examples of where men failed to be humans and sank to something much less:

- I had an engineer that insisted I sit whenever I spoke to him because a woman could not stand in his presence.
- I had an interview where the guy told me to meet him at his house, then we would go to the office but first I had to unload his dishwasher.
- I had an engineer who refused to report to me on the org chart even though he did in fact report to me.
- I had a Senior Vice President who told me that if I keep things up maybe one day, I could run a meeting because he didn't realize I was one of his managers.
- I had an interview where the guy asked me if I could cook biscuits then told me to bring a dozen when I reported to work on Monday.
- I had the Director that always had me take notes and order lunch regardless of if I was a senior member of the team, and the list goes on and on.
- The worst and one I remember the most vividly was, when I was told my raise would be one number, but it was less on my paycheck. When I asked about it, I was told that a male colleague needed more money, so he gave me less than we discussed, but it's okay since you have a husband to provide for your family. Yeah, you read that right.

You would be astonished by some of the things that have happened and been said to me over my career. I never had a female that I could speak with about any of these things, some male advisors just laughed at my stories, while my mentors were flabbergasted and ashamed to hear such things.

## 3.3 POWER ENGINEERING

I persevered and kept pushing, even more determined to climb the ladder. What things specifically did I do? In the first part of my career, I focused on technical knowledge. I went out to the field as often as I could so that I could learn how and why things worked. I asked questions and always went back to tell the person thank you for the information. Being thankful and showing gratitude is extremely important as you build your network.

Once I had a good technical foundation, I started to focus on my soft skills. I took speech classes, I attended leadership seminars, and I solicited feedback. This building of skills would lead to taking on projects in which I would lead a team.

I knew that financial and business acumen was the next item on my list, so I set out to get an MBA. I also transitioned to project management so that I could apply the principles to real-world examples. As my skills improved, I managed larger and larger projects and teams.

I climbed the ladder and I want other women to do the same thing, but the path is not always straight. Remember to own your career and make strategic moves to get you where you want and to the level you want to achieve.

I love power engineering. I'm passionate about a sustainable energy future. It is inspiring and energizing to me knowing that what I do will help to not only provide power to people today, but to provide clean power for generations to come. It's not just a job, it's my passion and that's why I love my career.

I was able to obtain the role of Director of Engineering for a Fortune 100 corporation. I was responsible for all their energy management and transmission solutions projects. While it was a dream job, I just couldn't stop thinking about starting my own company.

A few years prior, there was a rumor that I was going to start my own firm, but it wasn't true. The number of people who reached out to me to show their support made me think that maybe I could and should give it a try.

The final push I needed was provided by my employer. Like most companies, they look at projects as products pull through vehicles since they made the highest profit on their products. They didn't really care if a project had margin erosion as long as products were sold, and the client was happy. So, when it came to meeting diversity spend initiatives, the procurement team often got creative. They would hire sleeve companies, companies that have a woman or minority certification, which subcontracts out the work to a non-diverse company. There is zero value added since they are merely a pass through. There is one business owner in particular that has this business model. When I pushed back about it, I was told that they made 10% on $5MM without any liability or risk, so I should shut up. I was venting about this when a coworker challenged me. Instead of complaining about it, do something about it and open my own company. So, when the opportunity presented itself, I took the leap of faith and left my stable corporate position to start Select Power Systems ("SELECT").

## 3.4 #POWERTOMAKEADIFFERENCE

I'm the founder, President, and CEO of SELECT. Women-owned engineering firms are few and far between. We stand out because we are different. Yes, we provide quality designs, on time, on budget, and with exceptional customer service, but a lot of other companies do that too. Just look at our website and you will notice that the first two tabs don't tell you what we do, but it tells you who we are.

We believe that we have the #powertomakeadifference. When the power grid is secure, sustainable, affordable, and reliable, it improves people's lives. If you don't believe me, try living without using power for a few days. We are focused on modernizing this critical infrastructure. In addition, we are supporting projects to move more toward clean energy. It is important that we ensure we provide a sustainable approach that will reduce the waste, pollution, and inefficiencies of our current generation facilities.

We not only apply the #powertomakeadifference to our work, but also to our lives. We encourage and support our employees volunteering in their local communities. We focus on improving the areas where we work and live.

I founded SELECT with the principle of the Archway. It is the second tab of our website. You will see the archway is multi-colored because we understand that diversity is part of what makes the team stronger and better. Employees are our foundation. Quality and customer service are the columns. Safety and sustainability are the arches. Reputation is the keystone. We don't just put this on the website, but we lead and manage by these core values.

During the pandemic, I decided that I was not going to reduce salaries, furlough, or lay off a single employee. Utilization rates were dismal, so I had employees take company-paid training classes virtually. We lost money, but the commitment to our employees has fostered loyalty.

As a female electrical engineer in a male-dominated industry, I take diversity, inclusion, and belonging seriously. That's why we focus on finding women and minority talent from the recruiting phase. We also strive to support other women and minority-owned firms by subcontracting with them. There is more than enough work in the energy field, so we can be collaborative instead of being competitive. As the saying goes, a rising tide raises all ships.

I must take a moment to give major props to my husband. I remember telling him that I wanted to leave my job, liquidate all our assets, and raid our retirement plans, so that I could start my own business where I was certain I would not make any money for three years. He didn't even hesitate. He even quit his job so that he could help me get it set up and then returned to work to ensure we have enough money coming in so that we could eat. Talk about being supportive.

Still, it was and is the most difficult thing I have ever done in my career. I take my role seriously and it keeps me up at night knowing that my decisions can affect the employees' abilities to provide for their families. While SELECT has experienced tremendous growth and the future is bright, I still don't match my corporate salary yet. I prefer instead to continually reinvest in the company. I couldn't be more excited to lead SELECT through its next stage of growth. There is nothing more rewarding than seeing your baby grow and the blood, sweat, and tears pay off.

## 3.5 SUPPORTING OTHER WOMEN IN STEM

Does the lack of female role models and mentors bother me? Yes. That's why I make it a point to support other women in STEM. I carve out 1–2 hours a week to provide free coaching. I listen, I give advice, I give examples of similar circumstances I experienced, how I dealt with them, and the outcome of my approach, and I cheer them on. I didn't do everything the best way possible, but I did things the best way I knew how. If I can share these things with other women, I hope they can learn from my experiences. It is my passion to support, encourage, and help other women. I blog at LeanedOn.com to share my thoughts along the way. I hope that my story inspires you. As my husband tells people, living with me is like riding a rollercoaster. There are moments that are terrifying because you don't know what's coming next. But, as soon as that ride ends, you get back in the seat and go again because it is so much fun. I hope you climb on your own rollercoaster. Connect with me on LinkedIn, comment on a post on LeanedOn, or send me an email. I'm here for you, cheering you on every step of the way.

## ABOUT THE AUTHOR

**Heather Eason** has a bachelor's degree in electrical engineering, a master's degree in business administration, is a wife, mom, an entrepreneur, a CEO, and a board member. She is taking the road that once was not allowed, currently less traveled, and is blazing the trail for young females and minorities today who are interested in pursuing a STEM-centered career. She believes that we all have the #powertomakeadifference.

# 4 The Importance of Male Advocates for Women and Under-Represented Sects in STEM

*Matthew J. Walker*
Licensed Professional Engineer, United States of America

## CONTENTS



If you're reading this and are male – congratulations! You've recognized that we (yes, I am male as well) have a vital role to play in advocating for and getting more women – and minorities in general – into STEM fields.

> *"The best engineers, after all, are team players, adaptable, tenacious when faced with the most difficult challenges, and dedicated to an ethical obligation that puts the public first. ... The 11th century philosopher Algazel implied that engineering is noble and praiseworthy because it is subservient and indispensable to the public. If the nobility of the profession is tied to that subservience, and the demographics of the public are rapidly changing, why must we so arrogantly believe that we don't need to recalibrate our thinking toward diversity and inclusion?"*[1]

Arguments have been made that, "women should advocate for other women" (and similar for other minorities); but if we wait for each woman to assist one or two more over the span of their career in engineering, we will still be having this conversation 75 years from now. Men (particularly white men) are the majority; as such, we tend to drive the conversation; we are the overwhelming "chorus" voice in the room. So, if we want meaningful change to happen, we must be participants and not spectators.

Women bring a different perspective; by not including them, we leave out a multitude of possibilities (at least 50% of the possibilities!). As engineers, scientists, and technicians, we constantly look to ensure we have considered every possibility, every potential outcome or solution. That is, by definition, the job! If we exclude, or at least don't include, half the population, how can we be sure we truly have considered everything?

DOI: 10.1201/9781003336495-6

Similarly, we cannot simplify this to a "this starts in schools" or "this is an industry issue" debate. We must advocate for BOTH at the same time. It may sound like a large undertaking to do all at once, but it really isn't. The two feed each other: advocate in schools and there will be more candidates for the workforce; higher more diverse backgrounds in the workplace and there will be more role models and mentors to encourage those in school.

## 4.1 UNDERSTANDING YOUR IMPACT

I began my career as a military aviator and engineer by pursuing a degree in mechanical engineering at one of the US military service academies. These classes were meant to set the foundation for what we would experience in the field. While there wasn't a heavy female presence in the field, women who were already in the aviation industry were not barred from pursuing their dreams of becoming maintenance managers or engineers. Being in the field, I had rarely seen discrimination of women in STEM. I saw women who were already in aviation and showed interest in becoming maintenance managers or engineers, and I offered to bring them into the department but that's not really the same as "advocating."

It was during my tour as an engineering instructor and aviation representative back at my alma mater that I had my first experience seeing what women deal with, some on a daily basis. That was the first time I overheard two male instructors explaining to two female students that they (1) really didn't belong in a technical field like engineering because it was "dirty," and (2) didn't want aviation as a career because "you can't be a pilot and have a family as a woman." This was particularly upsetting for me, not only for the gross mischaracterization of both professions and the obvious disrespect to those two cadets, but neither of the instructors were aviators or Professional Engineers! They were speaking entirely from personal bias and assumption and offering it as guidance to future generations.

During that same tour, I had a female, senior engineering student ask for assistance. After asking her some questions to force her to think through the problem some more, I helped her look up the info she needed. Her response, to me, was overwhelming:

> Just wanted to say, you're probably the best instructor we've had in 4 years. Not for any reason other than you don't talk down to us, especially the girls, and you tell everyone to come to you if we need help, and when we do, you don't send us away with some snide comment – you actually stop what you're doing and help us.

I didn't feel like I was doing anything big or different than I would've wanted an instructor or friend to do for me. That's what we're talking about needing to do across the industry; if we say, "I want to help," then when people come to us, we need to Stop. And. Help.

## 4.2 BECOMING AN ADVOCATE

Many companies are focusing on Diversity, Equity, and Inclusion (DEI) right now, but some are in it for the wrong reason. DEI initiatives are a definite focal point in the news, and to not speak out is to be lumped in with those opposed, which can be bad for business. But if the talk doesn't match the actions, are you really working towards bringing more women, and minorities in general, into the profession, let alone your company?

Companies who say they are inclusive for women and minorities but seem to hire just to fill a particular percentage or ratio of the company staffing simply to say it's being done are exacerbating the problem rather than helping to solve it. Once the team is reduced to mere numbers, the actual issue at hand becomes lost: if the numbers are used solely to meet a Key Performance Indicator, then it undermines any true efforts for inclusion.

So, how do you know you're doing it for the right reasons?

1) Make it personal – You don't need to single-handedly change the face of our profession, of your company, or even your department. Simply start by helping one person, and it doesn't have to be fancy or heavy handed; as the age-old adage says, "think global, act local."

   If you know of a woman – or any under-represented member in engineering – in your immediate circle trying to get into engineering – encourage them! Help remove obstacles. Write a recommendation. Assist with networking: one of the areas that may be a challenge for many is with networking, as they may feel insecure or struggling with the pre-existing team cohesiveness (in some instances, this is similar to being the new kid on a sports team; in the worst examples, it's the "good ol' boys club" – whether real or perceived).
2) Volunteer your time – The most precious resource any one of us has is our time, and freely giving that to assist others demonstrates sincerity, commitment, and is a hallmark of good leadership – all things as engineers we should strive to be.
3) Learn to recognize when others are acting "in a discouraging manner" and step in. No "bystander syndrome" or expecting someone else to deal with inappropriate behavior; we have to hold each other accountable. Even if what was said was positive and included "all the right things," if the body language or unconscious actions give the appearance of favoring one applicant over the other, it says a great deal more about the underlying culture of that organization than all the platitudes and social media write-ups. Even small signs, such as not facing them or with your body oriented towards the door, can indicate you're not actively listening and looking for a way out of the conversation.
4) Organizational culture plays a tremendous role. Think about the culture within your team, work group, department, or company. Watch the interactions at the next industry day or symposium. What messages are being sent, intentionally & unintentionally, verbal and non-verbal? Think about your last expo event: is the booth staffed entirely with white males? If a man and woman (or any minority) walk up at the same time with questions, was one of those individuals focused on or ignored?

As for some basic dos and don'ts that anyone can start today:

DON'T:

1) Empathize/sympathize. If you're not a woman or minority, you don't know that they're feeling or what it's been like for them. Don't insult their struggle or their story by automatically trying to relate to it. Better yet, acknowledge that you don't know what that's like.
2) Feel like you must have an answer for every question. Like any good engineer, we don't memorize everything, but we know where to go to find what we need; so, if you don't know, say you don't know. Instead, steer them toward someone you know who might (another female engineer perhaps), provide an introduction, and offer to help.

DO:

1) Listen!! Not just to hear, but to understand, comprehend, and appreciate. You can't advocate effectively if you don't understand.
2) Be consistent in your messaging. That's not to say you can't change your approach as you go, but don't be too quick to assume you're not getting through. Watch your audience and recognize if they're not responding because they don't understand or because they're not motivated.
3) Be patient and persistent. The loudest voice in the room isn't the one that wins the argument. Change is hard and no matter how hard you fight or how good your approach, you are going to hit roadblocks. Instead, approach this like an engineer with a new task: attack the problem (not the person), re-evaluate any assumptions, test your plan…and "if at first you don't succeed, try, try again."

## 4.3 HANDLING IMPOSTER SYNDROME

With regards to roadblocks and persistence, there will be those who oppose any efforts you may make, to include attacking you personally or questioning your motivation (only you can know your true reasons for being an advocate). Sometimes, those comments can hit you so hard that it completely throws your focus. Some may even get so discouraged or turned off by those comments that we choose to stop advocating because of the negative reactions of those around us.

There will always be nay-sayers ("oh, you're only doing this because being a women's advocate is "in fashion" right now"). You didn't listen to the voices that said you couldn't do it, it was too tough, you'd never make it, and not to waste your time becoming an engineer when you first started, so why listen to them now?

Part of leadership is sticking up for those around you, creating paths for others that come after you. Advocating for women in STEM is doing exactly that.

Perhaps you are just "on the bandwagon," and perhaps you're not – only you know what your motivations are. And that's truly the only difference between being an advocate and an opportunist.

In the Spring 2021 edition of PE Magazine, professional engineer Zohaib Alvi wrote about dealing with imposter syndrome as a successful engineer. What he wrote also rings true for being an advocate within engineering.[2]

- "Talk to your peers, mentors, and bosses and make connections" – this goes both for you and for those you're trying to help bring into the profession. Networking isn't just for career advancement, but for support and advice as well. People are tribal; you need to build your tribe and help others build theirs.
- "Celebrate [your] accomplishments, large or small." It may seem small to you, but to others, it may represent a big step forward. And celebration doesn't have to mean cake and balloons. A simple, "Hey, great job!" or "That's awesome! Good work" can go a long way. Best compliment you may ever give or receive: "Thank you. I really appreciate everything you're doing around here." Don't be self-conscious (either of you) about giving and accepting credit when credit is due. Humility is good leadership, but you can be too humble. Besides, you never know who may read/see/hear about your success and be inspired to continue their own journey.
- "If you're in a position to mentor others, do so." Pay it forward; pass it on. Someone helped you get where you are (whether you're aware of it or not), and you owe it to return the favor. Pay it forward.

And this is how the change we desire will come to be. As American anthropologist Margaret Mead is often quoted, "Never doubt that a small group of thoughtful, committed, citizens can change the world. Indeed, it is the only thing that ever has."

## ABOUT THE AUTHOR

**Matthew Walker** is a 21-year veteran and licensed Professional Engineer. A Naval Aviator with multiple FAA licenses, he holds a BS and MS in Mechanical Engineering. His career includes responses to Hurricanes Katrina, Florence and Michael, the 2010 Haitian earthquake, and Deepwater Horizon disaster; teaching Thermodynamics, Aerodynamics, and Advanced Mathematics at a US military academy; and Chief Engineer, directing his service's airworthiness program. As his federal department's 2020 Engineer of the Year award winner, he was also a Top 10 Finalist for NSPE's Federal Engineer of the Year.

## NOTES

1 "Why Should I Care About Diversity in Engineering?", National Society of Professional Engineers' (NSPE) Diversity, Equity and Inclusion (DEI) Advisory Committee, PE Magazine, July/August 2020.
2 "Don't Let an 'Imposter' Derail Your Career," Zohaib Alvi, P.E., PE Magazine, Spring 2021.

# 5 STEM and Cars
## *A Collision Course*

*Becky Mueller*
Insurance Institute for Highway Safety,
United States of America

## CONTENTS



## 5.1 NAVIGATING THE COURSE

Smash! I was 10 years old watching an evening news segment on one of the first car crash tests from the Insurance Institute for Highway Safety (IIHS). My fascination with cars, desire to help people, and dream career collided that evening, as I listened to a crash test expert describing how crash testing cars can protect drivers and passengers.

I've always wanted to help people. In grade school, I wanted to be a medical doctor – helping sick or injured patients, saving lives. But I also had a growing fascination with cars; I can't really explain why. I loved all aspects of cars including their styles and performance capabilities. I spent hours sketching concept cars and reading car magazines. I dreamed about working at a car company – perhaps crafting clay models of concept cars or testing vehicle performance on a racetrack. I even created a cartoon book where I became the Chief Executive Officer of a car company and led the company from bankruptcy to success. Lofty goals for a 9-year-old!

Watching that first car crash test made quite an impact. Car crash testing was one way I could combine my love of cars with my life goal of saving people's lives. I began to dream about becoming a crash test expert.

Reality check. Having a dream that doesn't align with society's expectations was difficult. Every time an adult asked me "What do you want to be when you grow up?" my response – "I want to crash cars for safety ratings" – received a wide range of negative reactions, among them, "What got you interested in *that*?" or "What an *odd* choice for a pretty young lady."

School friends also questioned my interest in cars and science. I vividly remember a chat with a middle-school friend where she called my interests "weird" and recommended I should be more interested in fashion and other "girly" things. But the dream of becoming a crash test expert lit a fire within me and I was on a mission.

DOI: 10.1201/9781003336495-7 

Achieving my dream was going to require a plan and a little bit of luck. I needed to figure out the education and career path needed to obtain my dream job. Today, I advise prospective STEM students to Google their interests and possible careers to help create their own roadmap of steps toward the goal. The Internet contains details about educational requirements, potential career descriptions, and profiles of STEM professionals, with contact information for networking.

None of those resources were available when I was in grade school. My grand plan was to focus on getting good grades, especially in math and science classes, so that I could get into a good college to study engineering. Since many automotive designers and engineers were mechanical engineers, I picked that major and hoped it would work out.

## 5.2 EMERGING AS THE FRONTRUNNER

At the college level, many young women struggle with the unattainable goal of perfection. Engineering programs are less about perfection and more about class completion. I watched so many young women quit chasing STEM careers when they received their first B- or C, because they fall into the mindset that it's perfection or nothing. I watched female classmates abandon engineering if they couldn't achieve a 4.0 or even 3.5 grade point average (GPA), while male classmates with their 2.8 GPAs pressed on and graduated alongside me. I recognized I would sometimes struggle with the classes, but, motivated to get to my dream job, I always kept moving toward my degree, even if it meant dropping and repeating a class or accepting a C grade. Once I had my diploma, the GPA on my transcript would never matter again.

During my freshman year, I attended my first career fair, where I observed the long lines of candidates waiting to talk to recruiters from automakers, along with the stacks of resumes sitting on their tables. I would have to do more than just major in mechanical engineering to get an automaker's attention. To set me apart from other candidates, I needed to gain experience working with cars.

A friend introduced me to the Hybrid Vehicle Team, an extracurricular club that converted gas-powered vehicles to innovative, fuel-efficient, and eco-friendly hybrid designs. A vehicle-proving ground held annual competitions, and automakers often recruited the Hybrid Vehicle Team leaders. This was my chance to stand out. I spent every Tuesday, Wednesday, and Friday night working with the team, sacrificing time that could be spent socializing and relaxing with friends. I took leadership roles, first as the group's mechanical group leader, and then as the overall team leader. I coordinated the final vehicle design, co-authored a research paper, handled media interviews, and led the team through the seven-day competition at General Motor's Milford Proving Grounds to a coveted second-place award.

In a broad undergraduate field like mechanical engineering, there is little room for free time to take electives in specialty areas related to vehicles or biomechanics, so many people pursue master or doctoral studies to gain more knowledge in a specific area of interest. I believed a master's degree would set me apart from other candidates when it came to finding a position in car safety. I approached my faculty advisor for guidance, and he gave me an undergraduate research project that I summarized in a written thesis and presented to a faculty panel.

I was interested in attending Purdue University for a master's degree, but I knew that with a 3.3 GPA, I also needed a faculty connection for admission. My advisor knew a professor at Purdue and connected me for a summer research fellowship.

## 5.3 DEPARTING THE SAFE ZONES

Chance plays a role in every career success story, but success also depends on a person's ability to seize an opportunity. A pivotal moment in my career occurred at a conference during my fellowship. At the conference banquet, I sat down at a table alongside a General Motors (GM) engineering group manager. During the dinner, we struck up a conversation about my involvement with the Hybrid Vehicle Team. He was so impressed with my enthusiasm and knowledge that he slipped me a business card and indicated, "GM would love to have talent like you." The GM manager then told me to email him with my top area of interest because he had connections to other departments.

As I sat staring at a blank email, I knew this was my one shot, my moment to seize the opportunity (cue Eminem's "Lose Yourself"). I wrote "car safety" and hit send. Four months later, I received my formal internship acceptance letter, my assigned department was listed as "vehicle safety." I made it!

From Day One of my internship at GM, I hit the ground running, asking my mentors questions, gaining proficiency in software, and filling my downtime with reading research papers and safety reports. I spoke up and reached out to them, because I knew this was my one chance.

I invited various safety experts to lunch or coffee to hear stories of their career journeys, so I could learn from their experiences and plan out my own next steps. The head of safety hosted an annual intern luncheon, where the group of interns presented their summer projects. I had a conflict during the group luncheon, so I boldly requested a private luncheon with him. That hour became the foundation of a working friendship and mentorship that has lasted more than 15 years.

Graduate school was not easy for me, but I kept focusing on the end goal. Based on my internship experience, I changed my graduate research area to biomechanics. I lost funding, had to work as a part-time teaching assistant to pay for the first semester tuition, and landed on academic probation as a result of a sub-par grade in a math class.

Still, I persisted. As graduation and a full-time job at GM was within sight, the stock market crashed and GM filed for bankruptcy. Chance, as I learned, doesn't always work in my favor.

## 5.4 REACHING THE PODIUM

As graduation drew near, I applied to over 100 job positions with no promising options. I even applied to my dream job at IIHS, but they had no openings. I cried the whole night I got that rejection letter; it seemed like all hope of pursuing my dream career was gone.

Then, 3 months after graduation, I received a phone call from the Transportation Research Center, Inc. (TRC), a vehicle testing and proving ground. The manager had received my resume months prior during a hiring freeze, but was impressed and saved it. I knew very little about the job details, but I figured I'd plan my next steps once I started.

My first job at TRC was the perfect next step toward my dream job. I was working as a government contractor to the National Highway Traffic Safety Administration on future safety regulations, which was a different aspect of car safety than GM. I was back on track for my goal of crashing cars for safety ratings.

From my first day on the job, I spent my downtime getting up-to-speed on all things related to crash testing. I remember scrounging through storage boxes of old research papers to learn more about dummies, crash reconstructions, and testing. About 6 months into working at TRC, my boss sent me to a car safety conference in Washington, DC. Once again, chance was about to work in my favor.

At that conference, I met the IIHS staff member who I had previously contacted about a job opening. I approached him to introduce myself and thank him for his previous consideration, but I received a surprising response. He remembered my resume and immediately told me of a new job opening at IIHS he felt was perfect for me. A month later, I accepted the position, which would allow me to realize my long-time dream – research engineer.

Over the first 11 years at IIHS, I have coordinated the crash testing of more than 120 cars, and I've led the development and implementation of three new car crash test programs – each with the potential to save thousands of lives each year. My work has produced more than a dozen peer-reviewed research papers, and I've been invited to speak at technical conferences across the globe.

Six years into my career at IIHS, my dream finally came to fruition. As I watched the evening news, a segment on new IIHS car crash test results aired. This time, the IIHS expert they were interviewing was me.

## ABOUT THE AUTHOR

**Becky Mueller** is a graduate from the University of Wisconsin and Purdue University in Mechanical Engineering with a specialization in biomechanics. She worked as a contractor to the National Highway Traffic Safety Administration in 2009–2010 on projects related to future car safety regulations. She joined the Insurance Institute for Highway Safety in 2010 and is currently a Senior Research Engineer. Her projects include identifying road safety issues and developing new crash tests. In her free time, she figure skates, volunteers at nursing homes with her two therapy greyhounds, and travels around the world with her husband.

# 6 STEM, Creativity, and the Question of Gender

*Daniel Shorten*
Optimain Limited, United Kingdom

## CONTENTS



## 6.1 WHY SHOULD I READ THIS MALE PERSPECTIVE?

Whether we choose to admit it or not, there are still large gender gaps and bias within science, technology, engineering, and mathematics (STEM) fields. Embedded stereotypes present obstacles for women in STEM and continue to play a role in ongoing discrimination and the under-representation of women in STEM professions. One dominant stereotype is that boys are better at maths and science than girls, which studies show is simply not true. In addition, the sort of typical behaviors considered appropriate for careers in STEM such as objectivity and rationality are generally consistent with perceived male gender traits. Alternatively, women are seen as kinder, warmer, empathetic, less analytical, independent, and competitive and therefore less likely to have the qualities and personality characteristics needed to be successful scientists or engineers. These assumptions significantly influence the career choices and the retention of women in STEM fields. These stereotypes have been historically driven by the general gender stereotypes, which are now generally accepted as outmoded and potentially detrimental to the furtherance of innovation and to science in general.

My position as a man, a husband, and a father to a young woman at the start of her journey through education and the workplace, is that we must embrace the whole range of attributes that are exhibited by both men and women and to do so in the

DOI: 10.1201/9781003336495-8

knowledge that we increase our human capacity for innovation when the people we teach, train, and employ are free to behave according to their inherent skill set. One quote that resonates with me is,

> Don't let anyone rob you of your imagination, your creativity, or your curiosity. It's your place in the world; it's your life. Go on and do all you can with it and make it the life you want to live.
>
> – Mae Jemison, first African American woman astronaut in space

I think that many women might be surprised to hear that the views expressed herein are not in the minority, but the popular view. Only those who feel threatened by women in STEM would be likely to oppose. The reality is that it is now women themselves who must learn to drop their own stereotypes and see that equality is there for the taking and that STEM subjects are for them as are STEM careers. Careers that they can make their own without having to be anything other than professional and capable. In fact, I would go so far as to say that the very stereotypical female behavioural characteristics that we talk about are needed in society within STEM, within manufacturing and engineering.

## 6.2 MY OWN EXPERIENCES OF GENDER AND STEM AS A MAN IN THE LATER STAGES OF HIS CAREER

We often need to describe ourselves to validate our position. Whilst this should not always be the case, as in the statement of age in a job application, (Why does age prequalify or exclude an individual when being considered for a job?), it does serve in this instance to add some weight to my position as an advocate of women in STEM.

I am a man in his late fifties, I am an engineer qualified to degree level in mechanical engineering, tribology, and asset management. I am a company owner and a director. I am a husband to a successful wife and a father to a young girl who is just starting to consider which subjects she will opt for at the high school (*GCSEs for the UK contingent*).

I have worked predominantly in male-dominated engineering environments all my career, which is at the last count, just over 40 years. I started as a mechanical engineering apprentice and studied whilst working, meaning day release and night school. Moving through engineering test labs after the apprenticeship, where we tested lubricants to destruction in pumps, gearboxes, and industry standard test apparatus. It was only when I started to interact with the laboratory staff that I routinely interacted with girls. Being a lab technician or professional industrial chemist was clearly considered acceptable for girls in the late 1980s, though there were no girls at all in mechanical engineering and we did not even think to expect there might be.

As I progressed along my career path, the usual road map was to learn the trade in the oil test laboratory and then go out "on the road" as a lubrication engineer supporting the commercial teams in the field. At no stage did I work with any women either within lubrication engineering teams or the clients with whom we interacted. The only exception being large machine tool users who had their own laboratories like the Ford Engine plant at Halewood, in the UK, who employed a very talented lady as laboratory manager. She was very well respected, and unusually (in my limited experience), a woman who did not get on in life because she exhibited male characteristics, but because she was simply good at her job and did not allow anyone to intimidate her. This was the first time in my career that I can remember thinking that it might be good working for a woman, as her method of motivating people was not based upon power or threat, nor did she in any way use her physical attributes, she was simply a good manager, interested in all her colleagues, very capable as a chemist and an engineer and a pleasure to work with. She went on to do very well, though we lost touch as I moved on elsewhere within my own career.

I have to say this was not always my experience of women in engineering, as I certainly found that several persons, who shall remain nameless, moved through the ranks as a woman because they were better bullies than their male counterparts, not necessarily by their physical prowess but by their lack of sensitivity and singlemindedness, lack of compassion and cold calculating attitude to getting on in their career. These people rose quickly within organizations but did not last long as they moved frequently from job to job.

Personally, I believe that many women "feel" that they must have very male characteristics to succeed, mostly as a reaction to the perception that if they fail it will be "seen" as being directly related to their gender – which is of course nonsensical, but frankly completely understandable. In addition, I fear that this belief is subtly reinforced by weak men in the middle ranks of our corporations and engineering organizations, of which I have met many.

By means of a few illustrations, I refer the reader to a publication entitled "Invisible Women" by Caroline Criado Perez ISBN 978-1-784-70628-9. Winner of the 2019 Royal Society Insight Investment Science book prize, the only female author in a shortlist of 6, but chosen from a panel of 5 (three men and two women), the book is described as follows.

> Imagine a world where your phone is too big for your hand, where your doctor prescribes a drug that is wrong for your body, where in a car accident you are 47% more likely to be seriously injured, where every week the countless hours of work you do are not recognised or valued. If any of this sounds familiar, chances are that you're a woman.

Invisible Women shows us how, in a world largely built for and by men, we are systematically ignoring half of the population. It exposes the gender data gap – a gap in our knowledge that is at the root of perpetual, systemic discrimination against women, and that has created a pervasive but an invisible bias with a profound effect on women's lives. From government policy and medical research, to technology, workplaces, urban planning, and the media, Invisible Women reveals the biased data that excludes women.

Whilst this is not strictly a STEM issue, it does make fascinating reading about how women in the workplace are often unconsciously biased in a negative way.

Example 1 – New York philharmonic orchestra – hardly any female musicians until blind auditions were introduced in 1970 (p. 92). A significant increase in female orchestra members was evident once the hiring committee were prevented from seeing the applicant perform. *By the 1980's women began to make up 50% of new hires.*

Example 2 – Even though companies employ meritocracy techniques for performance evaluations, women are often accused of being aggressive, emotional, bossy, while men are not qualified negatively in this way. (p93).

Example 3 – Being published as an academic is a sure way of career progression, but evidence shows that women are significantly less likely to be published by institutions that do not employ blind academic auditions. As a result, women are cited less in academic papers (p. 96), and as such, this restricts career progression.

## 6.3 AS A PARENT OF A YOUNG GIRL WHO HAS YET TO MAKE LIFE CHOICES

She is 12 and curious. She is delightful and frustrating in often equal amounts. She has no idea what the next 10 years will bring, but by the time these years have elapsed, she will have most likely set the course for her adult life by the choices she will make. In general, this fact in of itself bothers me, as it is most likely that she will not make great decisions yet and I don't want her to find that opportunities are reduced, and doors are closed for her. She may change course and she should be able to do that. I feel this personally because I believe I did not navigate school well, made poor decisions, and as a result, had a significant number of options eliminated for me. As a result, I can honestly say that whilst I'm very happy where life's career paths have taken me, I still don't know what I want to do, but have a good idea of what I don't.

If my experience is to be considered unremarkable and familiar to many people, then adding to the mix the potential restriction of opportunities in STEM subjects will make our daughters, be unable to find their way into careers that stimulate and fulfill them.

I want my daughter to find something that she is truly passionate about and that might be in STEM, but it might not. Either way, I want her to have as many options as possible. Now in reality, it is much easier to get into STEM subjects and careers because there is a significant improvement in awareness and a hunger by learning and training establishments to attract more women into STEM. The problem it seems is in the deeply rooted cultural position that predominates where women don't yet see STEM careers as an option. Yet most, if not all the women, I have counseled on this subject who are in STEM positions choose that route because of their innate curiosity and a passion for problem solving. These are universal qualities! Furthermore, if anyone can find a career they adore, they will never have to work a day in their lives, so why restrict them?

Unfortunately, regarding STEM, girls begin to doubt themselves and their abilities at an early age, likely as they become more self-aware and unconsciously see science brilliance as a male condition. Consider asking a child to draw a scientist and you will most likely get a sketch of a man in a white coat with crazy hair.

We know instinctively at this point in history that there is a lack of representation of women STEM practitioners in society, and until we see a balance of male and female images, just as we might balance other biases, this is unlikely to change.

Consider this, the Norwegian volleyball team was fined when the ladies refused to wear bikini bottoms, which they felt unnecessarily sexualized them. This may not be an example directly related to STEM, but it does illustrate that we are continually affected by traditions that are significantly outdated and which then promote meaningless gender differences where gender is not actually a factor. It is my sincere belief that gender plays absolutely no part in the ability of any person to excel in STEM. Our biases and perspectives around what roles we have in society are at the root of this. This is not about active discrimination, far from it, but we have an imbalanced view of the value of gender in society.

## 6.4 INNOVATION IN DESIGN SHOULD NOT BE LIMITED BY GENDER

There are innumerable examples of designs that are clearly built from a male perspective.

- **Shaving Razors** – Originally designed for men but made pink to "attract" a wider market without ergonomic consideration.
- **Mammogram scanners** – My own experience is guided by my partner's experience of breast scanning and her view that the process could not have been designed by a woman, as the machinery may be perfectly functional, but it is not ergonomically aligned to the needs of the patient and can be very uncomfortable and painful.
- **Crash test dummies**. While men are more likely to be involved in a car crash than women, women are 47% more likely to be seriously injured than a man and 71% more likely to be moderately injured because of the physical differences between the genders. This is likely to be in no small way impacted by the fact that it is only since gender-specific crash test dummies being used in performance safety testing for car crash performance that these statistics have improved. Since the first crash test dummy in the 1950s, crash test dummies have been designed to represent a male in the 50th percentile. A female crash test dummy was first introduced in 2011, the inclusion of which in tests has sent some car safety star ratings plummeting.
- **Speech recognition software** is perhaps the artificial intelligence we meet most, on our phones, in our cars. However, these are much more likely to accurately recognize male speech than female speech. Mildly annoying perhaps, but an in-car voice recognition system designed to improve safety of using features while driving may in fact cause more dangerous situations if it fails to work correctly.
- **Personal protective equipment** (**PPE**). Women are often allocated smaller versions of male PPE. These are often uncomfortable, and not as effective when worn by women.

## 6.5 EMPOWERING WOMEN IN STEM IS NOT ABOUT FAIRNESS, IT'S ABOUT SURVIVAL

Very often when we talk about diversity, we are really talking about fairness and about eliminating prejudice. We do this because it is simply the right thing to do, where we differentiate about orientation, culture, race, size, age, etc. We need not ask "why?" but "Why not?".

In the workplace, this is often supported by guidance and regulation to outlaw prejudices that may lead to unfair and unjust discrimination. I personally feel injustice in my core when I see it first-hand or witness it in the news and online. I genuinely wonder what people fear and why they are mean and nasty to people just because they have a point of difference. I know it's almost human nature to gather in groups where we are similar, that's acceptable for me, but to actively discriminate or to turn away when presented with discriminatory behavior is thankfully, in my lifetime, becoming generally unacceptable. That said, it's still there and we need to avoid complacency lest it creep back in.

Active discrimination aside, I would suggest that the argument for a more balanced mix within the gender of STEM practitioners, is that we need to embrace the skill sets that predominate within each group of our people and build our innovation and human engineering resources around the simple understanding that to exclude certain skills and competence will ultimately lead to a decline in capability and a stagnation in human development. We must embrace all forms of diversity to set the individuals upon tasks to find novel solutions to the big questions and use STEM topics and our ability to be creative when addressing issues in all our societies. This is best said here,

> Sciences provide an understanding of a universal experience; Arts are a universal understanding of a personal experience… they are both a part of us and a manifestation of the same thing… the arts and sciences are avatars of human creativity.
>
> – Mae Jemison

## 6.6 WHY WE MUST EMBRACE GENDERLESS DEMAND IN STEM AND EMBRACE THE DIFFERENCES BETWEEN THE SEXES

The notion that certain subjects are for boys and others are for girls is outdated. We have soldiers on the front line, boxers, fighters, pilots, truck drivers, and a raft of other professions that would traditionally be considered as male occupations that can be quite admirably performed by anyone. Gender is a means by which to describe yourself, which is why we are also able for persons to define themselves in non-binary terms. It matters not to me what you call yourself but whether you are a decent human, a professional person who is capable and honorable, and your actions are noble. If we restrict by stereotype, persons who may have gone on to do great things, then surely, we are all the weaker as a result.

It is as stated earlier, it is our differences which give rise to innovative solutions and free us up from our blinkered approaches to problem solving. Similarly, if our only tool is a hammer, then we will only see problems as nails to be hammered in. The more tools we have at our disposal and the greater number of creative options we can access, then all problems become opportunities, work becomes rewarding, and the quality of life for everyone improves.

Difference is good, without which we stagnate and die, we must evolve and embrace new challenges and I look forward to the day when we no longer need to write about this because it is a thing of the past and consigned to the history books.

## ABOUT THE AUTHOR

By day, **Danny** is a lubrication and oil analysis professional and machinery health manager with experience of over 40 years. At other times, he is a husband and a father to an energetic and talented young teenage daughter. In what remains, he is a bass guitarist and dedicated provider of the groove! In 2016, Danny formed Optimain Limited, a UK listed company specializing in condition monitoring advisory services. He has developed and delivers training to operators and disparate teams who wish to develop their lubrication skills and become certified practitioners. He is an approved trainer for ISO 18436 pt4 lubricant analysis at Category III, ICML MLE, and a certified asset manager within the IAM.

# 7 Scientist, Engineer, Manager...Infertile

## *A Guide to Navigating Your STEM Career while Struggling with Infertility*

*Priya Santhanam*
Amazon, United States of America

**CONTENTS**



4.1 million
This is the number of women in STEM struggling with infertility[1, 2].

When we talk about the challenges facing women in STEM today, what comes to your mind? For me, it all boils down to two things: Recruitment and Retention. A critical element of retention that we seem to be missing as a global society is infertility. While there are a lot of policies established for parenthood and beyond, there is very little addressing the arduous path some of us have to go through to become parents in the first place. This can range from the inability to conceive to trouble sustaining the pregnancy or miscarriage. There are women around you in the workplace navigating this right now and suffering in silence. For couples wanting a biological child, it is a physically, mentally, and financially exhausting process to achieve that goal. Hundreds of needles, blood draws, ultrasounds, unexpected doctor's appointments, tons of shattered hopes, and disappointments go into creating that little human. All that while showing up at work, giving our best and continuing to kindle the fire propelling us toward the career goals we always wanted to accomplish! Indra Nooyi said it best – '*The career clock and biological clock will always be in conflict*'.

I find the dichotomy between these two entities fascinating today (and frustratingly ironic back in the years that I struggled with infertility!). Figure 7.1 illustrates a typical timeline in the life of a woman navigating a rigorous STEM education and

DOI: 10.1201/9781003336495-9

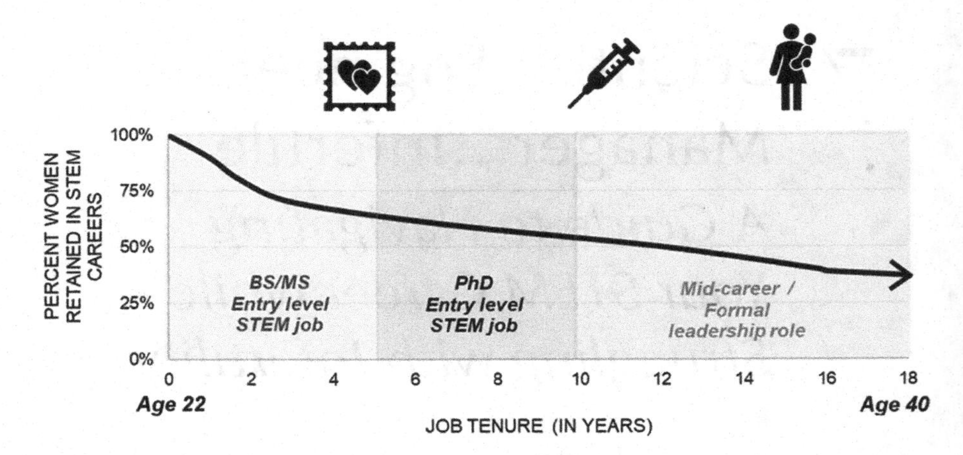


**FIGURE 7.1** Attrition of women in STEM jobs as a function of age/job tenure in STEM. [Data is juxtaposition along with the conflict in timelines between personal development (icons on top) and the career development (text) milestones].

career while simultaneously building her family life. Note that these are observational milestones and not prescriptive timelines! Observe how pivotal career and family moments in both these tracks occur at exactly the same time, leading to the attrition of women in STEM shown in the plot. Approximately 50% of the women entering STEM leave by the time they reach mid-career. Let that sink in – half of the women you started your career with are no longer in the workforce[3]. If this doesn't illustrate the severity of the retention issue for us, I don't know what will!

Of course, this entire attrition is not attributed to infertility alone. It truly takes a village to raise children and it takes a toll on any parent who has invested in their own career advancement just as much as they are in caring for their family. There is no data in literature today that helps identify what part of that attrition is from the path to motherhood and what can be attributed to the motherhood penalty itself[4]. However, even if the hypothetical contribution of infertility is a tiny 1%, we need to take care. That translates to a large enough absolute number that it is critical in this space where we have a lack of sufficient female representation.

Recently, a couple of countries of the world grabbed attention for the right reasons – they established policies around miscarriage, which is a common issue for women struggling with infertility. Here are some excerpts to provide perspective on what this looks like:

- The Australian federal government introduced legislation that will add miscarriage to the compassionate and bereavement leave entitlement, which means 2 days of paid leave will be provided to those who miscarry before 20 weeks[5].
- New Zealand's parliament approved legislation that offers 3 days of bereavement leave from work to mothers and their partners if they go through a miscarriage[6].

Two days after my first miscarriage, I showed up at work. I was walking away from a meeting when a co-worker made this comment: "*Hey, I heard you were sick and had to miss that important meeting with operations yesterday...and here you are! Were you really sick or just wanted to escape the meeting?!* (wink!)". My colleague meant this in good humor. However, my physical and emotional state was so raw that it took all my energy to walk away politely to shut myself in my office and cry. I recount this incident to give you a peek into what it's like to show up at work after a major incident like this happens in life. Contextually, 48–72 hours of paid leave post miscarriage seems like a joke to me. However, we absolutely cannot deny that this is progress. It is a recognition of the issue, the need to establish policies and a great first step in what will hopefully evolve into a meaningful outcome.

> Interestingly, in India, the maternity benefit act of 1961[7] entitles women to 6 weeks of paid time off, as long as they can provide proof of miscarriage. However, the stigma surrounding miscarriage is so high that I am not sure if this policy gets used as much as it needs to be.

All said and done, I have no control over these policies in my current career choice. Neither is it a constructive exercise for us to explore the depths of the role of Governments at this point. Instead, let's shift our focus on tangible things that you and I can do to add value in this arena for ourselves, friends, family, and colleagues. I will share my infertility story in Section 7.1 and be as candid with you as possible on exactly how this impacted my career advancement in Section 7.2. In Section 7.3, we will highlight what you can do to help yourself as a woman experiencing infertility. Section 7.4 will discuss the role you can play as a team member and a manager to support your colleague experiencing infertility.

I would like to believe that I live a life of privilege. As an immigrant in the US who came from a typical middle-class family in India, my family worked very, very hard to get me where I am today! Nonetheless, it is a life where various resources are available to me. Despite this privilege, it was extremely challenging figuring out how to sustain a thriving career while creating a family life that I desperately craved for. I hope this narrative gives you a transparent view of what infertility entails, how it can make or break a woman's career and why we all need to care about making the noise around this topic.

## 7.1 HUMBLE BEGINNINGS

*A PhD in engineering from an American university, 4.0 GPA, 15 conference presentations, six peer reviewed journal publications, multiple outstanding awards at my school and community, a job at a prestigious company willing to sponsor my permanent residency paperwork.* This is the profile I graduated with in Fall 2013. By Fall 2015, I suffered multiple miscarriages and learned the hard way that I was infertile. "Priya Santhanam, Infertile" became my identity in my own head and "Priya Santhanam, PhD" did not seem to hold any meaning.

In India, it is common for the elder women in the extended family to ask the younger women "*when is the good news?*". What they truly meant is to ask me when

I was going to tell them I was pregnant. Every time I was asked this question, it ate away more of my pride in my academic and professional accomplishments. In hindsight, did I handle this difficult time in my life well? Absolutely not! Was this a natural reaction to an episode like this? A resounding, "Yes!"

I still remember the very first time I miscarried. It was 7:50 a.m. on a beautiful day in 2014. I was on my way to work, all pumped for an important presentation I was to make at 9 a.m. that day. I parked my car and realized I was bleeding right before I got off. Believe it or not, my first thought was "I have that stakeholder meeting today!". Messed up priorities in my head! I called my boss and was bawling. I think I said something like "I just miscarried and need to go to the hospital." I was crying so bad, I have no idea to this day how he made sense of my words! His response was "Oh Gosh! Go to the hospital…do you need me to drive you?". Instantly, he became a superhero in my head.

Once it was all done and dusted, I went to the doctor's office with so many questions swimming in my head. How do we troubleshoot this? What timelines are we looking at? What tangible actions are we going to take? And I got nothing! I was told this happens to a lot of women and I am not considered infertile until I have three miscarriages. Now, you can imagine how frustrating it can be to go through all that and hear someone diminish your experience. Subsequently, it happened again. This time, I had to collect the fetus sample myself at home 1 day after the miscarriage and put it in a box to send for lab analysis. It broke me inside out. My mind was cloudy. I think my IQ instantly dropped 40 points during that phase! My husband had the forethought and determination to fight for me with my doctor and contest the "3 miscarriages" rule in America. He refused to believe it and let me go through it again and reached out to insurance and made an appointment with a reproductive endocrinologist. If you live in America and had a similar experience, please contest it. You hold the power to choose what you decide to put your body through.

From this point on, it is a standard "multiple IUI's-IVF-needles-ultrasounds-blood tests" infertility story that I will not bore you with! The critical part is that I never talked about my journey with anyone. One day, in a moment of vulnerability (right after my first IVF cycle), I shared it with one of my close friends and she immediately put me in touch with her friend going through the exact same thing. I was shocked! "WHAT?!! There are other STEM folks this happens to?". After that point, slowly but steadily, I built a tiny community of female infertility warriors at my workplace. I found five other women going through this, and we often talked about the joys and pains of infertility in one-on-one conversations. This ranged from "what do I tell my team when I get called for last minute ultrasounds because my ovaries don't understand meeting schedules?" to "My back hurts from needles and I can't sit down today!!".

I am so grateful for my tribe, and hope everyone finds at least one person they can relate to and share their story without being judged. Despite the ups and downs, I love my adventure. It is unique, it is my own, and most importantly….it gave me "My Why" shown in Figure 7.2.

**FIGURE 7.2** "My Why"; [My two children, both conceived via in-vitro fertilization (IVF)].

## 7.2 AMBITION CALIBRATION

So how exactly did infertility impact my career? I'm sure you will agree that no one can predict how life's events choose to unfold and how certain career opportunities lead to a certain elevation. Like a true engineer and a scientist, with lack of data, I turn to the narrative!

In 2013, here are the top two goals I wanted to achieve in my 5-year career plan:

a) Get an operations assignment to gain business exposure.
b) Bring that knowledge back to the research organization I was in and become a leader.

There was a direct impact of the stories I highlighted in Part I based on my performance in 2014–2015. I failed to deliver the best as my usual self and received an unfavorable performance review. As a result, neither of my two goals materialized as shown in Figure 7.3. While this felt like a huge setback to me at the time, in hindsight, it was quite alright.

Despite not receiving a full-blown operations assignment, around 2016, an incredible job came my way to support an operations team for 2 months. The company called these "suitcase assignments" where you traveled Monday to Saturday, spent a couple of days back home, and then flew back to the operations site overseas. When I was told about it, I did not even stop and consider this for a brief second. I immediately said NO. I was in the process of going through my injections as part of the in-vitro fertilization (IVF) treatment and figured it would be too hard to do both.

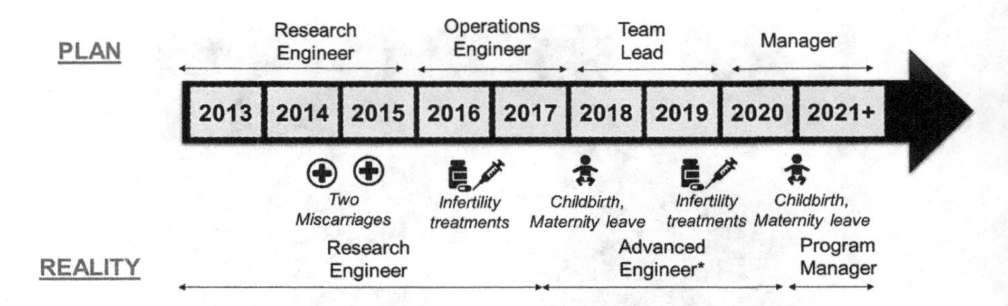


**FIGURE 7.3** Career plan vs. reality influenced by an unexpected infertility interruption. [Timeline showing milestone plans by years vs. the milestones, which occurred in reality].

Now, you may ask – how critical were these two items for my career advancement and did it matter that I didn't get these? Perhaps it didn't matter in the grand scheme of things. I was lucky to get the opportunity to support operations later in my career (after becoming a mother) and also lead the technology development for a pretty cool technology. It all came – just later than I wanted it. Life has its own way of giving back. (I think the "luck factor" has a role to play too. I truly detest that word, so maybe we'll call it "right place at the right time" instead!) I'm thankful for the prospects that came my way and this may not be true for everyone. Infertility can have more significant impacts on career advancement for women than it did in my case, the more extreme impact being to quit a growing career.

## 7.3 HELP YOURSELF

If you are a woman experiencing infertility, what are the tangible things you can do to help your own career advancement?

1. **Share your journey and seek help at work (if you feel comfortable doing so!)**
   One of the biggest regrets I have is not talking about my infertility earlier on. It would have shattered the feeling of isolation and get the help needed to do more and better at work. Despite common perceptions, sharing a vulnerable moment can be seen as a sign of strength in the workplace and your colleagues may be more willing to do everything in their power to help you get through it than you think! There will always be those who walk around with unconscious or conscious biases about women in the workplace and our ability to perform in light of biological factors – we can't change them overnight! *However, sharing our journeys is a first step in helping address this bias.*

   After I declined the suitcase assignment mentioned above, I met a colleague who did exactly what I had said no to. She was going through IVF herself and got a similar position to be in Germany on a suitcase assignment. She administered the shots to herself and shared a YouTube video that beautifully demonstrated how to go about doing this. I wish I had shared with her earlier to get that valuable tip that might have changed my career trajectory! Or rather, I wish I had the grit and determination she did in that difficult moment to be able to prioritize both career and personal life.

2. **Plan in advance and block out all appointments on your calendar**
   One of the biggest challenges while going through infertility treatments is the unpredictability of the multiple lab tests and doctor appointments. The first time I went through it, it felt incredibly overwhelming and I truly struggled balancing my work meetings and the innumerable doctor's appointments. At times, I would go for an ultrasound in the morning and get a call to come back late in the evening to check how the ovaries responded within that period … all that with a packed workday of meetings and deliverables!

   Over time, I got better at planning my days and weeks around these appointments. At one of my appointments, I spent time with the nurse and asked her help to chart out a Day 1 through N generic plan for a typical cycle. This was an incredibly insightful thing to do as it allowed me to block certain days in my week where I expected to be called in unexpectedly. As long as I could predict these at least 2–3 weeks in advance, I could avoid critical meetings being scheduled on these days and also complete deliverables ahead of critical timelines and release them at the stipulated timelines.
3. **Create a career roadmap and adjust it as events unfold**
   While going through the infertility process, the overwhelming feeling of "not doing enough" in your career is common. One way to get past this is to have clarity on your career goals and aspirations. Invest a couple of hours in chalking out a 5- and 10-year plan for yourself. Instead of focusing on job titles (so, unlike what I did at the time!), focus on skills you want to grow and what span of influence you want to have a fulfilling career. As your own infertility expedition evolves, calibrate the plan and adjust for changes. Make it as realistic as possible so you can take pride in a sense of accomplishment when you do achieve the milestones you set out for yourself!
4. **Allow yourself space and grace**
   If the message did not come through so far, allow me to say it explicitly now: Infertility is HARD. It can make you question your entire framework of life, starting from fundamental beliefs and all the way to your own self-worth. I wish someone had given me the advice to be kind to myself, not only allowing the space to wallow, but also appreciate my own body and mental strength for what I was going through.

   Try to visualize the dichotomy of these situations: Making the most important presentation of your career on the same day that you are bleeding from a miscarriage. Getting that dreaded call from the doctor's office that tells you why you cannot get pregnant and then walking into a team social event where everyone else seems to be having fun. Your body and mind hurting from all the needles that have touched your body and climbing into equipment to perform inspections at operations. These are tough career moments as a woman in STEM. They teach us a critical life skill – resilience. However, that knowledge comes much, much later! In the moments that you are going through this, allow yourself a day off or reprioritization of deliverables with your manager or a cupcake or a spa day – embrace whatever form self-care takes for you.

Space and grace!

## 7.4 HELP YOUR COLLEAGUE

If your team member gathered the courage to share with you that they are experiencing infertility, what can you do to help them?

1. **Respect their privacy**
   If you only do one thing for your colleague, it must be this – respect their privacy and do not share this around to encourage office gossip. The information could have come your way intentionally (from a place of trust from your colleague) or unintentionally (a vulnerable moment where they felt the urge to share). Irrespective of the intent, you now have a critical responsibility to protect that information and ensure they feel safe that you were confided in.
2. **Be an ally**
   Being an ally is a difficult job. Particularly if you have not personally experienced whatever your colleague is going through. One practical way you can demonstrate your ally ship is as follows: If you can explicitly see your colleague is exhausted or feeling overwhelmed, help them prioritize their work and identify areas where you or another team member can help. Try to be as informal and constructive about it as you can be. The more specific you can be in helping them discover ways to not compromise their value addition in the team while pacing themselves appropriately, the bigger the win!
3. **If you are a manager, place trust and provide flexibility**
   For those in a formal leadership role, the best thing you can do is to provide flexibility and show them you trust them. Coming to you asking for workarounds for four doctor's appointments every week can be draining for both parties. Create a circle of trust for your report to build their own schedule that allows them to contribute to the team's goals while working around their own personal situation. If performance issues emerge, have the discussion just as you might for any other employee – with empathy, data, and clarity in the path forward. One way to go above and beyond can be to help them build a near and medium-term career roadmap outlined in Part III. Brainstorming together can truly help build a robust plan, and your unique position as a manager can help your employee be realistic about the roles and skills that work well for them.

I share my career aspirations and the associated upsets unabashedly here with one sole intent – to help you see that it is okay if things do not happen perfectly at the time we plan them. It is okay to prioritize whatever is critical at the time to make your life feel fulfilling from a holistic perspective. There is no right answer to an ideal career path.

Your career is not a race! In my career, I've seen colleagues (male and female) who have been able to take on and run with certain opportunities when I was unable to do so. I've seen jobs I craved for ending up in other's laps and felt defeated about it. Years later, I've seen these individuals now grappling with the exact same complexities of creating and caring for a human at a time when my life allows for a bit more flexibility from a practical perspective to take on those challenges as my kids grow. Everyone's journey is unique – embrace your own!

I love this quote by Marie Curie, as it exemplifies everything I hope every woman in a rigorous career like STEM can embrace: "*Nothing in life is to be feared. It is only to be understood*".

## ABOUT THE AUTHOR

**Priya Santhanam**, PhD, is an engineer by training, scientist at heart, Manager by day, and writer at night. She grew up in various parts of India and moved to the U.S. to pursue her MS and PhD in chemical engineering. Priya has published over 10 scientific papers, holds three patents, and has presented at over 20 technical conferences. She is a mom of two young children and is constantly hustling to find that coveted personal life – work-life balance! She is a STEM advocate and currently leads an initiative to bring awareness around infertility in the STEM community via her blog www.scientistengineermom.org.

## REFERENCES

1. https://www.statista.com/statistics/1116527/share-women-stem-country/
2. https://www.who.int/health-topics/infertility#tab=tab_1
3. Jennifer L. Glass, Sharon Sassler, Yael Levitte, and Katherine M. Michelmore, "What's So Special about STEM? A Comparison of Women's Retention in STEM and Professional Occupations", *Soc Forces*. 2013; 92(2): 723–756.
4. https://www.aauw.org/issues/equity/motherhood/
5. https://www.fairwork.gov.au/leave/maternity-and-parental-leave/pregnant-employee-entitlements
6. https://www.employment.govt.nz/about/news-and-updates/bereavement-leave-to-cover-miscarriage-stillbirth/
7. https://labour.gov.in/sites/default/files/TheMaternityBenefitAct1961.pdf

# 8 STEM Women's Career Challenges and Possible Solutions

*Shalini Aggarwal*
IIT Bombay, Mumbai, India

*Beauty Kumari*
Asian Development Research Institute, Patna, India

*Harihar Jaishree Subrahmaniam*
Aarhus University, Denmark

*Subathra Rajendran*
Rapid Serviz, India

*Anand Swaroop*
White Hat Jr, India

*Vrinda Nair*
Concordia University, Quebec, Canada

## CONTENTS



DOI: 10.1201/9781003336495-10

## 8.1 ABSTRACT

*"Not all heroes wear capes. Some wear lab coats. We need more women in science to beat pandemics and solve the most challenging problems we face today"*. Science, Technology, Engineering, Mathematics (STEM) is the engine of the economy. It is where innovation occurs, and creativity is allowed, a place where one can make a difference. STEM domains are traditionally perceived as male dominant. It gives the basic perception of scientists as socially awkward, glasses-wearing men and can easily influence young minds incorrectly. A relatable mentor for young minds can play a significant role in dispelling stereotypes to achieve gender parity in STEM. Recently, many organizations have helped women achieve their STEM dreams: by vocalizing and supporting their needs, providing networking and skill development opportunities, offering mentorship, and more.

The COVID-19 pandemic has heightened the challenges women face in the STEM workforce. "*The worst choice that you can force somebody to make is to pick between their family and their career,*" bestselling author and futurist Jacob Morgan said. Eventually, this is the case for many women in their mid-careers who compromise their careers for family. Lockdowns worldwide left female academics with increased domestic responsibilities, leaving less time for taking care of work and, more importantly, themselves. The deepening crisis relating to women's mental health in STEM in pandemic times needs urgent attention.

## 8.2 INTRODUCTION

Over the last 50 years, women have made enormous strides in academics and the corporate world. With the expanding importance of STEM education becoming recognized worldwide, it has become essential to encourage women in STEM. The intricacy of evaluating the extent to which scholarly activity may be classified as STEM education is exacerbated by multiple opinions on what it means to be a STEM educator. STEM involves thinking critically, logically, and analytically. It also requires approaching problems persistently and creatively. Thus, STEM subjects and activities can teach youth how to think differently by sufficiently understanding their targeted disciplines. To diversify STEM disciplines, we must confront the preconceptions and biases that still exist in our society.

On the one hand, women have made significant progress in traditionally masculine sectors such as business and law. Women's educational advances in scientific fields, on the other hand, have been less remarkable, and their advancement in the workplace has been slower. Why are so few women becoming scientists and engineers in an era when women are becoming prominent in medicine, law, and business-like areas? Scientists and engineers are created in colleges and institutions, but the groundwork for a STEM career is established early in life. We must address this bias and try to transform the system to promote young girls' involvement in STEM, be it at school or home.

Furthermore, girls and women can be encouraged to pursue careers in STEM by paying close attention to the atmosphere in the classrooms, workplaces, and across the culture. Women in STEM fields may help in boosting innovation, creativity, and competitiveness by adding a different perspective to the field. Many organizations have been working to help young women achieve their STEM career dreams by vocalizing and supporting their needs, providing networking and skill development opportunities, offering mentorship on pursuing a career in STEM, and more. Additionally, recruiting, retaining, and advancing women, conducting awareness campaigns, competitions, training, and projects for girls/young women to engender interest in the STEM fields can be addressed through organizational methods. Scientific and technology products, services, and solutions are more likely to be better created and represent all users with a more diverse staff. So, it is imperative to increase the representation of women in science, technology, engineering, and mathematics.

## 8.3 CURRENT CHALLENGES AND POSSIBLE WAYS TO IMPROVE THEM

The lack of women in STEM hinders the growth of the field due to scarcity in mindset diversity and new perspectives. One of the significant side effects of not having a comfortable environment for women in STEM may also lead to brain drain in India. Brain drain is the nation's loss of highly skilled professionals through emigration to developed countries. India needs to turn "Brain Drain" into "Brain Gain." A country should be able to attract the best talent worldwide by eliminating defects such as corruption, nepotism, red-tapism, gender biases, and many more factors, one by one, to make the environment suitable for dwelling a fertile and beautiful mind. However, India's existing policies or programs cater to several issues like career break, relocation, research opportunities associated with foreign institutions, and awareness generation among girls on a STEM career. It has somewhere failed to tackle or deal with the critical and unavoidable causes, which pressurize women to trade-off between career and family life. The minuscule status of women in STEM is not only because of the lack of interest, as this gender represents 43% of the total STEM enrolments in the country, but unfortunately also, it falls to 3% and 6% for Ph.D. in Science and Engineering, respectively, as per the AISHE, 2018–19 report[1]. Hence, the problem is more complicated, involving multi-dimensional issues like uncertainty to finish Ph.D./significant time commitment, cloudy job prospects, working through the night, work-life balance, especially after marriage, salary, flexibility at work, motherhood, and more. The challenges posed to women in their careers can be divided into three categories based on the place in their career timeline, such as early career challenges, challenges while working in the field, and rejoining post-career gap.

### 8.3.1 Early Career Challenges

#### 8.3.1.1 Mentoring

Mentoring youth has a positive effect on students' growth and academic success. Having a mentor can sometimes make all the difference when deciding on a professional path or a topic of study. Mentoring is crucial in assisting young people in exploring and developing their interest in STEM subjects. Mentorship is a two-way communication since it develops a transferable skill. A strong connection between a mentor and a mentee is beneficial for both, as it will help them learn and experience. This interactive approach is a great way to acquire and help one to gain leadership skills and a sense of achievement in helping others.

Additionally, mentoring is one lever used to activate to advance more women in their fields, to help them gain access to capital and economic opportunities they might otherwise miss, and to be better prepared for opportunities when they come. All of us – mentees and mentors – are women in the making or already boldly declared to be in the sisterhood. The support of each other is required at a fundamental level, which goes beyond mentoring and even beyond sponsorship.

#### 8.3.1.2 Volunteering

Volunteer work encourages the participants to gain and grow their network and perspective. One tends to learn teamwork and achieve results with a diverse group of people. The extensive research on personalities and coordinating with other team members who work in different research fields can help ignite many qualities and opportunities. The different platforms to volunteer, such as 1 million women in STEM (1MWIS), Anukarniya, Astrobiology Graduate (AbGrad) community, Simply Neuroscience, and many more, are openly accessible for the youth to learn and grow rich.

#### 8.3.1.3 Journal Clubs and Internships

Discussing and evaluating literature reviews is an excellent exercise if one joins a journal club. Meeting peers working in diverse research groups and investigating findings is fun and engaging. The more you learn, the more you discover. Periodically, differences in opinion equally promote a contrasting style of thinking. One also learns problem-solving and how to broaden skills. Research evolves every day, and staying updated has become an important notion. Additionally, small internship programs such as the New York Academy of Sciences, the junior academy[2] in early career, might help one ignite the will of fire for knowledge and gain new perspectives of approaching problems.

#### 8.3.1.4 Extracurricular Activities

Learning new strategies and soft skills helps boost knowledge inside and outside one's work life. It makes up the understanding of unique concepts with a more comprehensive grasp on the subject. It is also estimated that around 30% of the extracurricular activities are considered for application evaluation. Extracurricular activities provide such training that helps gain competitive skills and makes one capable of having something different.

#### 8.3.1.5 Ask Yourself as Many Questions as Possible

The motivation to start a new task happens due to the pipeline set by the society, inspiration obtained by watching some extraordinary performer of the field, or self-drive. However, for one to keep moving ahead, it needs constant motivation, which is impossible if the reason behind "why we started it" is clear to us. As Simon Sinek's book explains, the importance and effect of knowing "why" on the success rate of the task being performed or planned are highly influential[3]. Hence, it's always advisable to ask yourself as many questions as possible when you are preparing to take the next step so that the efficiency and chances of success are in your favor.

#### 8.3.1.6 Strength and Weakness Analysis of Oneself

To succeed in life, one must understand their strengths and weaknesses well. Knowing strength increases the chance to succeed in the task at hand, while knowing weakness will reduce the suffering and chances of failure due to ignorance or false notions about self. Hence, one of the best ways to move forward is by strength, weakness, opportunities, and threats (SWOT) analysis before starting a task. This helps you to analyze all the possible scenarios and prepares you for the worst situation that may arise. Once the SWOT analysis is done, the person may start working on the targets required to complete the task at hand. It will help achieve the goal when the real job begins; as mentioned by David Goggin in his book, "*one doesn't rise to up to their expectations but falls to their level of training*."[4]

#### 8.3.1.7 Awareness about Scholarships and Initiatives

Money is also another significant impediment in higher education that may arise due to the family's economic condition or deeply rooted mindset of old marriage traditions for girls' marriage. This results in the lack of monetary support for girls' education. Hence, awareness about the scholarships and sponsors available for the target field might help one get rid of the scarcity of finances required to acquire higher education. STEM college scholarships are awards that focus on STEM fields. Students have to dedicate themselves to science, technology, engineering, or math to qualify for these scholarships, increasing the odds they will contribute to the workforce in those areas (**Table 8.1**). The scholarships mentioned in the table represent the tip of the iceberg; there are many more fellowships that one may avail and fulfill their dreams. Few of the consortia of the scholarship information includes Anukarniya[5], wemakescholars[6], Biotechnika[7], and Lotus STEMM[8].

## TABLE 8.1
## List of Scholarships for Girls in Early-Career STEM Careers

| S.no. | A. Scholarships for Women in STEM | Field | References |
|---|---|---|---|
| 1 | ABC Humane Wildlife Control & Prevention, Inc. Academic Scholarship | STEM | https://abcwildlife.com/about-us/women-in-stem-scholarship/ |
| 2 | British Council scholarships for women in STEM | STEM | https://www.britishcouncil.in/study-uk/scholarships/womeninstem-scholarships |
| 3 | BHW Scholarship | STEM | https://thebhwgroup.com/scholarship |
| 4 | AWIS Kirsten R. Lorentzen Award | Science | https://www.spsnational.org/scholarships/lorentzen |
| 5 | Aysen Tunca Memorial Scholarship | Science | https://www.spsnational.org/scholarships/tunca |
| 6 | Women Techmakers Scholars Program | Technology | https://www.womentechmakers.com/initiatives |
| 7 | Society of Women Engineers (SWE) Scholarship | Engineer | https://scholarships.swe.org/applications/login.asp |
| 8 | Abel Visiting Scholar Program | Mathematics | https://www.mathunion.org/cwm/resources/funding-scholarships |

| S.no. | B. General scholarships in STEM | Field | References |
|---|---|---|---|
| 1 | AIAA Foundation Undergraduate Scholarships Program (Multiple Opportunities) | STEM | https://www.aiaa.org/get-involved/students-educators/scholarships-graduate-awards |
| 2 | Microsoft Scholarship Program | STEM | https://microsoft.recsolu.com/external/requisitions/28_5r8UnqXBHK1Cf_pNfcw |
| 3 | USRA Scholarship Award (Multiple Opportunities) | Science & Engineer | https://www.usra.edu/educational-activities-and-opportunities/usra-distinguished-undergraduate-awards |
| 4 | ASDSO Scholarship | Technology and Engineer | https://damsafety.org/apply-scholarship |
| 5 | Herbert Levy Memorial Scholarship | Science | https://www.spsnational.org/scholarships/levy |
| 6 | Peggy Dixon Scholarship | Science | https://www.spsnational.org/scholarships/dixon |
| 7 | Scholarship America Dream Award (Multiple Opportunities) | Science | https://scholarshipamerica.org/dream-award/ |
| 8 | ASME Scholarships Program (Multiple Opportunities) | Engineer | https://www.asme.org/asme-programs/students-and-faculty/scholarships/asme-scholarships-how-to-apply |
| 9 | BMW/SAE Engineering Scholarship | Engineer | https://www.sae.org/participate/scholarships/bmw-sae-engineering-scholarship |
| 10 | Steinman Scholarship | Engineer | https://www.nspe.org/resources/students/scholarships/steinman-scholarship |

### 8.3.2 Challenges While Working in the Field

#### 8.3.2.1 Know Your Rights and Be Safe

Male-dominated industries and occupations are particularly vulnerable to reinforcing masculine stereotypes, adding challenges to women's lives, knowingly or unknowingly. Some of the manifestations of these challenges can be in the form of societal expectations and lack of beliefs about women's leadership abilities, pervasive stereotypes, higher stress and anxiety, lack of mentoring and career development opportunities, and sexual harassment. These hindrances can be present in any field, irrespective of gender, time, and place. Women use various coping mechanisms to work in male-dominated work environments, such as distancing themselves from colleagues, accepting masculine cultural norms, and acting like "one of the boys," which exacerbates the problem by contributing to the normalization of this culture and leaving the industry. However, quitting or ignoring may not be helpful in the long run to eliminate these incidents. Hence, one should be aware of the rights and policies imposed by the institutes against such discriminations and reach out to the concerned authorities. Generally, these authorities or help cells are by the name, "women cell." If a women's cell is not present, one may reach out to independent women cells.

#### 8.3.2.2 Work-Life Balance

Balancing the demands of office life and home life can be overwhelming for any working parent. More than 75% of all caregivers are female, meaning that in addition to juggling the demands of work, a large portion of women are also juggling the demands of caring for a loved one. Women's unpaid labor outside of the workforce often results in a career break that has both short-term and long-term effects. "*The worst choice that you can force somebody to make is to pick between their family and their career,*" bestselling author and futurist Jacob Morgan say. "I think that's terrible." As the coronavirus pandemic continues to force millions of employees to work from home, experts predict these flexible work options could be here to stay. The work-from-home period will prove that employees can still be productive when you allow them to work outside the office. "Forward-thinking organizations" will implement long-term, flexible work programs after the pandemic, allowing parents, especially women, to work schedules that accommodate both their personal and professional demands. "Any leader who practices empathy will say, 'Look, you can have both,' "You don't need to sacrifice one for the other."

### 8.3.3 Rejoining Post Any Career Gap

Another challenging phase in a woman's career struggles is resuming her career post a maternity gap. Women are highly motivated to restart their careers after fulfilling their role as a mother, but this also leads to a gap in knowledge due to continuously growing fields. However, the gaps can be easily covered if given a guiding, positive environment and opportunities. Major corporate companies listed below understand the career comeback for women professionals and provide opportunities. One of the significant losses due to this is the example of Dorothy Reed, in 1902, when she discovered a way to differentiate between Hodgkin and non-Hodgkin lymphoma, but she couldn't continue her work due to childbirth and had to leave the institute. Later, the lymphoma classification took around 50 years to come to light that could have been way faster if Dorothy had been allowed to continue her work. However, with time, things have turned for the betterment and for encouraging women in STEM, and there are some excellent initiatives taken by different companies and governments (**Table 8.2**). Furthermore, the Indian government's adoption of policies to encourage and retain women in STEM is not new. For instance, Women Scientists Scheme (WSS) was introduced way back in 2002 to promote gender equity in science and technology by providing opportunities to women to re-enter research after a career break.

Lately, the government has introduced other policies like KIRAN, CURIE, BioCARe, etc., to prevent the leakage in STEM fields. Yet, the percentage of women in STEM remained at 14%, an improvement of only 3% between 2015 and 2019[9], UNESCO Institute for Statistics (UIS) reports. Moreover, this improvement is not substantial compared to other Asian countries like Azerbaijan, Thailand, Georgia, and others (World Economic Forum, 2019)[10].

The current policies and schemes are applaudable but do not resolve the root cause that prevents women from continuing in the STEM field. For instance, a program like

**TABLE 8.2**
**List of Scholarships for Women for Post Gap Career Opportunities**

| S.no. | Company | Name of Initiatives | Reference |
|---|---|---|---|
| 1 | Accenture | Career Reboot for Women | https://www.accenture.com/in-en/careers/local/career-reboot |
| 2 | Google | gCareer | https://india.googleblog.com/search/label/gcareer |
| 3 | GE India | RESTART | https://www.ge.com/sustainability/ourphilanthropy |
| 4 | Capgemini | Relaunch@capgemini | https://www.capgemini.com/gb-en/careers/relaunchcapgemini/ |
| 5 | Goldman Sachs | Returnship | https://www.goldmansachs.com/careers/professionals/returnship/ |
| 6 | Axis Bank | Re-connect | https://www.axisbank.com/docs/default-source/CSR-Reports-and-Disclosures/people.pdf?sfvrsn=0 |
| 7 | Godrej | Careers 2.0 | https://www.godrejcareers.com/ |
| 8 | Hindustan Unilever | Career By Choice | https://careers.unilever.com/ |
| 9 | TATA | Second Career Internship Programme (SCIP) | https://www.tatasecondcareer.com/#/ |
| 10 | IBM | Bring Her Back Program | https://www.ibm.com/employment/techreentry/ |
| 11 | Phillips | Back in the Game (B.I.G.) | https://www.careers.philips.com/apac/en |
| 12 | Intel | Home to Office | https://www.intel.in/content/www/in/en/products/docs/boards-kits/nuc/nuc-home-office.html |
| 13 | Department of Science and Technology | Gender Advancement for Transforming Institutions (GATI) | https://dst.gov.in/scientific-programmes/scientific-engineering-research/women-scientists-programs |

KIRAN and BioCARe provides an opportunity to re-enter research. Still, none of the policies discusses creating an environment that can help them continue their job without dropping out. Similarly, none of the policies talks about the flexibility to work from home or discourages late-night work in laboratories. Additionally, after investing 12–15 years in academics from undergraduate to Ph.D., one is not secure to get a well-paying job. The crèche policy is also not followed by labs or other research institutes because of fewer researchers. Hence, along with the opportunities created by various government and private bodies, a nurturing and guiding environment is required (**Box 8.1**) to avoid the loss of women's power in STEM.

**BOX 8.1 POSSIBLE POLICIES TO RETAIN BRAIN DRAIN OF WOMEN POWER FROM STEM**

Overall, in addition to providing the opportunity to re-enter or enroll in STEM, it is important for the government or institute to provide an enabling environment to women, which will encourage them to stick to STEM by overcoming various challenges. The following points present potential steps that the Indian government or policymakers can adopt while designing a women-friendly STEM field policy.

1. Providing research opportunities at the undergraduate and postgraduate levels will intrigue research interest among students: Similar to western universities, Indian government should make conducting research mandatory for professors, increasing the opportunities for students to work as research assistants/associates under the research project and hone their research skills. Besides, it will also make use of professors' talent for the country's R&D.
2. Private and the government should incentivize public research laboratories/hospitals and others to release fellowships/internship/training/volunteer for recent graduates or undergraduates: Similar to the Centre for disease control and prevention in the USA or National Health Scheme in the UK, the Indian government should also incentive research institutes, laboratories, and others to launch opportunities for students from middle and high school, higher educations, and others. For example, Bharat Biotech can tie-up with the University of Delhi or other universities, giving the flavor of real-life research to students and benefiting the country out of the agreement.
3. Childcare facilities should be made available on campus: The Maternity Benefit Amendment Act, 2017, makes it mandatory to have a crèche if the organization has 50 or more employees, but because of fewer women in laboratories or research institutes, it does not work. The government needs to re-evaluate this.
4. Provide flexible Ph.D. or postdoctoral positions where women can work from home or take flexible working time: Like other countries, part-time or flexible Ph.D. or postdoctoral options will encourage women to stick to STEM while balancing social pressure and other responsibilities.

## 8.4 WAYS TO IMPROVE ONESELF AND TO BECOME EFFICIENT

In pursuit of becoming the best in one's field, for generations, we humans have been exploring various ways to become efficient and to do so, one way is time management. Time management is a skill that one needs to learn to experience a fuller life. On the contrary, some might think living in the moment will lead to a happier life. When time management skills are developed, one starts noticing how to get more things done and still have more time to do something that was missed out otherwise, having more freedom than ever. The unconscious time-wasting actions will be a thing of the past.

Technically speaking, time management does not mean managing your time. The time cannot be controlled. Time transcends physicality; it is eternal. Time runs the same for every one of us. In reality, what you are managing is your attention. You have a limited amount of attention; you can give your full attention to one thing at a time. When you order your attention, you consciously choose what to focus on rather than unconsciously allowing your attention to wander away. There are a lot of time management techniques you can learn; some of the very effective methods, which are a must in your tool kit are as follows.

### 8.4.1 Parkinson's Law

Parkinson's law states that work expands to fit the time you have to complete the job. To take an example – If you have 5 days to prepare for a test, you will take 5 entire days to revise and prepare for the test. But now, for the same test, you have 2 days to prepare, then you put out more intensive preparation and complete the same syllabus in only 2 days.

Steps to apply Parkinson' Law

1. Write down the task you have at hand.
2. Decide the amount of time you need to complete the job and write it down.
3. It will help to write the start time and expected finish time of your task.
4. Try to complete it by the mentioned time; if not, readjust to more realistic timelines.
5. Repeat step 1 to start another task.

### 8.4.2 Pomodoro Technique

*Did you ever feel like your brain is not processing information when you are reading?* It generally happens when you study for prolonged hours. You know you have to start working on something, but you want to delay it. You procrastinate. You wonder why it happens that sometimes you will be so inefficient at learning. According to scientists, an average man has an attention span of about 25 minutes, after which the ability of his brain to grasp information decreases, that is, brain efficiency decreases. For the first 25 minutes, brain efficiency is well above 95%, and after that, it drops rapidly and goes as low as 10%. Therefore, studying for prolonged hours is not recommended.

Scientists also found that relaxing from your task for 5 minutes will bump your brain efficiency back to 95%. Hence, it is an excellent practice to take a 5-minute break after every 25-minute study session. (*Note: 25 min is for an average person, you can increase time up to 50 min, based on your attention span*)

The Pomodoro technique is developed to work at high efficiency most of the time. In that way, one can complete more work in less time. The phrase "study smart" definitely fits here.

Steps in a Pomodoro technique

1. Decide a task you will work on
2. Set the Pomodoro timer (anywhere between 25 and 50 minutes)
3. Work on the task till the timer goes off (don't allow yourself to get distracted)
4. Take a break (anywhere between 5 and 15 minutes)
5. Go back to step 2
6. After completing the task, go to step 1
7. After repeating four Pomodoro sessions, take an extended 30-minute break.
8. Repeat step 1 to step 8 until you are done.

### 8.4.3 Pareto Principle (80/20 – Rule)

Pareto Principle or 80/20 Rule says that 80% of all the consequences come from 20% of the causes. It means that 80% of the results come from 20% of the effort, and to get the remaining 20% of results, you need to put in 80% of the effort. For example, 80% of the revenue of any company comes from 20% of its customers. These customers are its core users, and the rest of 20% of revenue comes from 80% of customers.

For some mysterious reasons, this Pareto Principle applies to almost everything in the world. It may not be precisely an 80/20 split; it can happen by 60/40, 70/30, 90/10, and 99/1 split.

Maximum desired output can come from a few causes, and your job is to find them and focus your effort toward them to get the most results with the least effort and less time.

Steps to apply Pareto Principle

1. List down all the tasks you have to do.
2. Arrange all the tasks in the order of priority.
3. You have to decide the priority in terms of urgency and the impact it can have once it is done
4. Select top 20% or 30% of the tasks and focus on completing them Pareto principles say that you get 70–80% of the results you expect to see if you can meet these tasks and ignore the rest.

It is beneficial to save time; instead of focusing on many things that take up a big chunk of your time, you will save tons of time and have peace of mind when you find out the critical tasks and finish them.

## 8.5 ALWAYS BE READY TO FACE FAILURES

Are you ready to embrace failure? Until now, we read about the possible hurdles and ways to overcome these hurdles, became aware of the opportunities available, and how to be efficient in self-management, but this only prepares us for the things in our control. No one can ever predict 100% of the scenarios that might go wrong. Hence, it is always advisable to be prepared for the failures with an open mind and keep moving forward – failure in obtaining admission in that dream school or that grant or that paper rejection. The list goes on! Often it becomes challenging to handle rejections, and one may take it negatively, self-loathing and question every decision. Admitting failure can be especially hard, where there is a culture of highlighting successes into a logical narrative to our scientific talks and papers. It's hard to remember that very often, and the entire story is not even told. Each successful experiment, each grant won, is a product of a million failures and, embedded in them, a million lessons. Each loss gives us something essential: a lesson, some new previously unknown information, and an opportunity to improve. So how to handle failures??

### 8.5.1 Recognize That You Are Not Unique

Academia does a great job of hiding the complex parts and glorifies the successes. But it's important to remember you are not alone in the failures encountered. You are not the first one, and you certainly won't be the last. For that grant application, which was not accepted, check the percentage of awardees, then inverse the number. Those are the applicants you are in company with. Far more applications are rejected than accepted. Accept it, reflect upon it and move on.

### 8.5.2 It's Not Personal

Unless you killed the reviewer's partner, burned their house down, or ran over their dog, the critique is about your work, not about you. Look at it from their perspective: the reviewers often have a limited time to make decisions reading a multitude of papers. They are bound to choose the ones that strike them the most from the pool of applications.

### 8.5.3 There Will Be Other Opportunities

Failure is not the end of the line; it is just a means to pivot. Grants roll cyclically, and many times, it's on the second or third time that the application is successful. There are a million journals out there, and your paper will find a home. Successful academics are the ones who are persistent and determined. Think about how many failed applications and missed opportunities you had to make it so far, and apply that same spirit to move on and find better opportunities.

### 8.5.4 Mindset Matters

Never take failures as a blow to your self-esteem; look at it as an opportunity to learn. It doesn't hurt to think about why you encountered failure. Don't think: "stupid reviewers, why didn't they understand me"? Instead, think, "*how can I be better so that even the reviewer from other fields can understand me perfectly.*"

### 8.5.5 Learn from Your Successes

What is seldom discussed is that success takes up a lot to work; it's one of the downsides. One has to put in the hard work to succeed. The idea that got funded, you had to put in the effort that made the application really good. You then move on to bigger and better views. At least, if you are unsuccessful, you can salvage the good parts and make a better application for the next round! Failure is a chance to reflect upon your methods and get even better at them. It's also a chance to learn about YOU, your strengths, resilience, and determination to go forward, no matter what life throws in your direction. So, embrace it; it might just be the best thing that could have happened to you.

## 8.6 CONCLUSION AND FUTURE PERSPECTIVE

Across STEM disciplines, there are generally far more men than women. Women's representation in science and engineering should continue to rise at the graduate level and again in the transfer to the workplace. Women have made significant progress in science and engineering, but they still make up a small percentage of the workforce in many sectors. If workers should make their workplaces more inclusive to people of all genders, leaders should raise awareness of female achievers in STEM fields. When it is said and done, prejudice is still a genuine issue that must be tackled. Discouraging so many bright minds like Dorothy Reed comes at a high price, not just for the people but for humankind as a whole. For the women in STEM workforces, the recent pandemic highlighted some of the difficulties they faced with remote work and increased caregiving responsibilities, severely hindering their career progressions. Several avenues have recently come up to tackle this problem by providing a roadmap for improving women's recruitment, retention, and advancement in STEM. However, efforts in this avenue must go beyond employing women and include creating a support structure within the workplaces to sustain women through their STEM journeys. There is also a need for more communities such as Anukarniya[5], WomHub[11], Lotus STEMM[8], MENSA[12], and 1MWIS[13], dedicated to providing opportunities to women by building networks, mentorship, and leadership opportunities for advancing women's STEM careers. Despite the deep-seated difficulties, there are hints of development. Belief in one's ability to thrive in a STEM profession is vital, but it is not the only determinant in developing an interest in a STEM career.

Whether you're a woman or not, choosing a STEM career course has definite benefits worth considering. These are some reasons to choose a STEM career, in particular, which can entice women on the fence about their options, such as job security, monetary benefit, advanced environment, handy skills, market gap, and status. STEM enables one to work in a high-tech world and spend days on the cutting edge. It's all about looking ahead, whether you're an aerospace engineer, a software developer, or a math instructor. The internet and a greater understanding of our culture are also due to STEM professionals. At their heart, STEM jobs all require solving problems and the skills to do so. Collaboration, logic, and analytics are all critical in STEM. Overall, technical skills are needed, as well as interest, independence, and imagination. While routine jobs are vulnerable to automation, STEM jobs are relatively secure. STEM graduates are more adaptable to evolving work requirements and are more likely to fit into newly developed positions. Although no job is guaranteed, STEM graduates are almost always hired right away for both men and women.

Additionally, it is also observed that the people with a bachelor's degree in a STEM field earn approximately 50% more than those with a Ph.D. in other areas. A biomedical engineer earns €66k on average, a data analyst €77k, and a mathematician €92k in the Netherlands. There are positions open, and more are being added all the time. As a result, there will be less competition. Every year, nearly 3 million STEM-related posts go unfilled due to a lack of eligible candidates. We need a massive force coming in supporting and highlighting women in STEM. A much better representation is the need of the hour.

While the number of women in STEM workforces still has some room to grow, global efforts to create significant changes to the status quo have a great promise for an equitable tomorrow. Developing positive STEM identities and relating to role models can help girls and women choose and pursue a STEM career. Several efforts in providing mentorship and leadership platforms have been made globally to dismantle women's misconceptions in choosing a STEM career. This may help in providing a gender-sensitive environment and institutional support. Although there is a long road ahead to achieve gender equity in STEM, empowering girls and women to develop their STEM talents to become future leaders is vital to provide a more robust and safer world as we meet the global challenges of tomorrow.

## ABOUT THE AUTHORS

1. **Beauty Kumari** is a motivated development professional with several years of experience working with NGOs across India. She holds an LSE degree in Global Health Policy. She is a staunch supporter of evidence-based policymaking in the pursuit of a just society, and as a result, she formed the AADER foundation, a section-8 company in India, dedicated to reaching the final mile through research and policy advocacy. Her organization focuses on environmental, health, and education issues. She is proud to have received a Commonwealth Shared Scholarship at LSE and to have received a Gold Medal during her post-graduation at XISS, Ranchi.
2. **Harihar Jaishree Subrahmaniam** is a plant evolutionary ecologist currently working as an MSCA Individual Fellow at Aarhus University. Apart from being a researcher, she is also a science communicator finding effective ways of communicating scientific research to the non-technical public. She represents the Marie Curie Alumni Association by being an open science policy advisor. She also mentors young girls and women to help them find a fulfilling STEM career. To help create a positive nurturing environment for

researchers, she founded a mental health support initiative for researchers all over the world to help them in their journeys in STEM careers.

3. **Subathra Rajendran** is an Entrepreneur and Freelance process Engineer. She did her graduation in Chemical Engineering, is a university rank holder, and soon after graduation was appointed as Scientist Engineer in Indian space Research Organization (ISRO), Sriharikota, where she was involved in solid propellant mixing unit for GSLV MK 3 project, which forms the first stage fuel for Space Launch Vehicles. Working few years in Sriharikota, she quit ISRO and joined with non-renewable (O&G) energy industry and has an experience of 17 years. She is very happy to be a woman of force in this male-oriented industry and break rules of gaining versatile experience in all aspects of onshore and offshore, petrochemical plants, including engineering design, procurement, commissioning, and operation. Being a woman despite of all social, cultural, and gender challenges, she managed to work in different parts of globe in India, South East Asia (Malaysia, Singapore, Indonesia and Thailand), Middle East (Dubai, Sharjah and Oman) Europe (Netherlands, Italy and France), Central Asia (Kazakhstan), and Africa (Congo). During this pandemic, she has strongly emphasized on the importance of environmental conservation and made a startup company working on renewable energy (solar), E-Vehicles energy storage devices (Li Ion batteries for E-vehicles and Inverters).
4. **Anand Swaroop** was born in 1993 in Hyderabad, state of Telangana. He completed his master's degree from IIT Bombay, and during this period, he discovered his love for books. The knowledge he gained from reading great books combined with his own experience of competitive exams led him to become an amateur writer and a mentor. He is now working as a senior manager in a prominent ed-tech organization. Even in his tight schedule, he makes time to help aspirants of competitive exams to give back to society. His favorite snack is a dairy milk chocolate; usually, he is a generous guy, but one won't find him sharing chocolate unless one forces it out of him.
5. **Vrinda Nair** is an aspiring scientist and is currently pursuing her Ph.D. in Physics at Concordia University, Montreal, Canada. She has an engineering background in biotechnology. Currently, she also serves as the treasurer in the FGSA at the American Physical Society. She was awarded Young Investigator Award by the AABS for her work in developing bio-colours for bioplastic. Vrinda is a published author-poet. She is also a certified professional life coach, solopreneur, instructional designer, violinist, artist, and digital illustrator. She actively supports many causes like women in STEM, supporting sci-artist and has worked with various organizations and offered volunteering services.
6. **Shalini Aggarwal** is a Ph.D. scholar at IIT Bombay. She has done her B.Sc. (H) in Microbiology from Delhi University. She qualified IIT-JAM with all India rank 40 and got admission to IIT Roorkee, scholarship-basis from the Department of Biotechnology. She secured a Ph.D. position at IIT Bombay

with the highest score in the entrance exam. She has received various awards such as a travel grant for poster presentation at Paris, merit awards at BSc level, newton Bhabha fellowship, 2020, and many more. She has two patents on infectious diseases projects and multiple publications, including research articles, review articles, and book chapters. She is a reviewer in a peer-reviewed journal, Nature's communication biology. Apart from academics, she is a taekwondo and chess athlete. Also, a proud member of MENSA India since 2010 – an international organization of high IQ people worldwide.

## LIST OF ABBREVIATIONS

**STEM** Science, Technology, Engineering, Mathematics
**MWIS** 1 million women in STEM
**AbGrad** Astrobiology Graduate
**SWOT** Strength, Weakness, Opportunities, and Threats
**WSS** Women Scientists Scheme
**UIS** UNESCO Institute for Statistics

## REFERENCES

1 AISHE Report 2018-19 | Government of India, Ministry of Education https://www.education.gov.in/en/aishe-report-2018-19 (accessed 2021-11-30).
2 The Junior Academy https://www.nyas.org/programs/global-stem-alliance/the-junior-academy/ (accessed 2021-11-29).
3 Start With Why How Great Leaders Inspire Everyone To Take Action Simon Sinek: Buy Start With Why How Great Leaders Inspire Everyone To Take Action Simon Sinek by SIMON SINEK at Low Price in India | Flipkart.com https://www.flipkart.com/start-great-leaders-inspire-everyone-take-action-simon-sinek/p/itm2c9c3b4b5a6ff?pid=RBKG59PDMHGTVWY3&lid=LSTRBKG59PDMHGTVWY3VVNVDF&marketplace=FLIPKART&cmpid=content_regionalbooks_8003060057_u_8965229628_gmc_pla&tgi=sem,1,G,11214002,u,,,395332127672,,,,c,,,,,,,&ef_id=Cj0KCQiA7oyNBhDiARIsADtGRZbnh8tzP9iN-YcuvMpY457agtsu30MGaKg0wF6yG-zDQ2MbTCD4TzUaAlcUEALw_wcB:G:s&s_kwcid=AL!739!3!395332127672!!!u!295092701166!&gclid=Cj0KCQiA7oyNBhDiARIsADtGRZbnh8tzP9iN-YcuvMpY457agtsu30MGaKg0wF6yG-zDQ2MbTCD4TzUaAlcUEALw_wcB&gclsrc=aw.ds (accessed 2021-11-29).
4 Goggins, D. Can't Hurt Me Self-Discipline, Mental Toughness, and Hard Work, *Bk Book Store*, 2018.
5 Aggarwal, G. Anukarniya https://www.buzzsprout.com/1255607/episodes (accessed 2021-11-29).
6 #1 Platform to fund your Study Abroad – Scholarships, Education loan|WeMakeScholars https://www.wemakescholars.com/(accessed 2021-11-29).
7 BioTecNika – Helps You Make Career in BioSciences Industry https://www.biotecnika.org/(accessed 2021-11-29).
8 About https://www.lotusstemm.org/about (accessed 2021-11-30).

9 Huyer, S. Gender Equality Will Encourage New Solutions and Expand the Scope of Research; It Should Be Considered a Priority by All If the Global Community Is Serious about Reaching the next Set of Development Goals. 20.
10 Gender equality in STEM is possible. These countries prove it https://www.weforum.org/agenda/2019/03/gender-equality-in-stem-is-possible/ (accessed 2021-11-30).
11 WomHub https://www.womhub.com/ (accessed 2021-11-30).
12 Barhate, H. About TMNP, *Tribal Mensa*.
13 Home | Million STEM https://www.1mwis.com/ (accessed 2021-11-30).

# 9 Collaboration Is the Next Generation of Innovation

## *How Working Together Can Be the Gateway to Creating a More Inclusive, Innovative Tomorrow*

*Shadrach Stephens*
Re.engineer, United States of America

**CONTENTS**



In this day and age, collaboration is a critical workflow component in many industries. As a business leader and an engineer by discipline, I would say that the most important lesson I have learned throughout my career is the power of collaboration and the impact it has on the development of the professionals around the world. Collaboration is the next generation of innovation, and in order for professionals to thrive in the 21st century, we all must engage. Reid Hoffman, one of the co-founders and an executive chairman of LinkedIn once said, 'No matter how brilliant your mind or strategy, if you're playing a solo game, you'll always lose out to a team'. It's quite refreshing that one of the leaders of arguably the largest professional based social media platform in the world believes that collaboration is the fuel behind success.

Even though areas such as science, technology, engineering and mathematics (STEM) are incredibly competitive, a collaborative environment can give a team the right impetus to stay on the cutting edge. In high-tech, fast paced environments, we can also strengthen professional relationships, which can pave the way to personal growth. Collaborating with others is a great way to get new ideas and gain a fresh

DOI: 10.1201/9781003336495-11

perspective that might enhance productivity and success. The traditional relationship between leader and individual contributor is quickly shifting. Today, most employees aren't just expected to complete a task. Their respective leaders increasingly value their input and everyone else, enabling a more collaborative environment that's more open to alternative solutions, change and innovations. Organizations that do not embrace a collaborative attitude might find themselves stuck in their comfort zone and unable to keep up with the times and their competitors.

So is collaboration > competition? Not many people know this, but Lamborghini was founded as a result of revenge of a tractor owner who was insulted by Enzy Ferrari, the founder of Ferrari. Lamborghini became frustrated with problems he had with the clutch in his Ferrari. He then went to visit Ferrari. Ferrari said, 'The problem is not with the car but with the driver!' and went on to advise him to look after his tractors instead. Today, Lamborghini sells more than half of what Ferrari does. Imagine this story if that conversation was one of collaboration and not competition. Maybe Lamborghini and Ferrari could have come together to produce some of the industry's most dynamic and powerful vehicles.

Our competitive nature is a key element of being human and respectful competition is not all bad. However, competition, on the one hand, can prevent us from achieving our greatest potential because it can be divisive. When we only think in terms of competition, there can only be one winner. Collaboration, on the other hand, is all about progressing as a whole. There is no winner unless the entire group benefits.

## 9.1 BETTER RESOURCE MANAGEMENT AND ENHANCED TECHNOLOGY MAKES COLLABORATION EASIER

Fostering a more collaborative environment can encourage teams and organizations to stay on top of their resources and manage their time more efficiently. This is also true when it comes to space organization in the workplace. A collaborative physical environment can lead to efficient office redesign, as companies rethink their professional environment in a quest to make the most out of them. Collaboration can make people feel more valued and engaged, and in turn, they'll use up their time and resources in a more productive way. By creating favourable circumstances to a collaborative environment, it is possible to kick-start productivity and create a link of connected ideas, resulting in innovation.

Some people hesitate to establish a more collaborative environment because they might think the new set up would require a massive logistic undertaking. However, modern professional tools enable collaboration to a whole new level. Today's tech-driven world has many ups and downs. However, one of the ups is undoubtedly how it is so much easier to collaborate with others and achieve seamless and excellent communication. Tools such as Slack, Clubhouse and Microsoft Teams are only some of the many options available to teams and organizations looking to embrace a more collaborative working relationship. Today's flexible digital technology allows collaborators to seamlessly exchange data and information. In addition to that, they can use such platforms to provide guidance and advice and keep a constant and clear line of communication with everyone involved.

## 9.2 IMPROVING COLLABORATION CULTURE THROUGH DIVERSITY OF THOUGHT

A collaborative environment is often defined by mutual respect between co-workers. Encouraging this strength is a great way to strengthen professional relationships and build stronger networks. In turn, company culture can also improve dramatically, making for a one-of-a-kind opportunity to foster a more substantial, cohesive and more effective team.

Diversity is being invited to the party; Inclusion is being asked to dance; Collaboration is choosing the music together, is a phrase that I use often, which adds perspective to how we should approach creating collaborative environments. Organizations that lead in inclusion and diversity (I&D) reap the diversity dividend because there are benefits of having a diverse and collaborative team. A diverse team improves the quality of decision-making and increases customer insight and innovation. An encouraging and inclusive workplace enables companies to hire and retain the best team. In fact, research shows that companies suffer a penalty for opting out of I&D. Those in the bottom quartile for both gender and ethnic diversity are 29% more likely to under-perform on profitability. Exclusion comes at a cost for those companies who are not proactively seeking engagement from everyone. Data shows that only 30% of the American workforce feels actively engaged, which leaves a massive 70% feeling disengaged. The cost of that disengagement to the U.S. economy alone is $450–$550 billion annually.

Many companies are making a valiant effort to drive an increase in their percentage of diverse leaders, which is very encouraging; however, looking beneath the surface at diversity amongst the lower levels of organizations, we see a trend that also needs improving. In many cases, the attrition rate for minority employees is two to three times higher than that of non-minority employees. Just as important, not everyone feels as if they belong or can bring their true selves into the work environment. We also know that there are pockets of our population who don't believe they can express themselves as they are while still collaborating and developing their talents.

The apparent opportunity is how do we drive collaborative engagement, reduce attrition rates among minorities, and also harvest the untapped, potential value for our organizations? There are three key indicators from the Work Institute 2019 Retention report that set the stage on where the opportunities are: (1) the number one category of employee retention, which directly relates to Career Development, has decreased by 32% since 2013; (2) advancement or promotional opportunities has decreased by 46% since 2010; and (3) lack of growth and development opportunities continues to trend up over the years, with a 170% increase since 2010.

## 9.3 B.R.I.D.G.E. THE GAPS

Twenty years ago, I started my first internship, so you can say that I have been in industry for just about two decades. In that time, I can count on one hand how many times someone asked me specifically about what they can do to support minorities in STEM or how we can partner to take the conversation deeper. Yes, I've been in many town hall meetings and group discussions, but I am directly referring to having one on one conversations about these topics. If I am honest with myself, I have actually never even prepared a response to such questions. I have thought about them but never in an articulate perspective that would lead to action.

As the world begun to take notice after George Floyd's murder, I received numerous invites to share my thoughts on I&D, and my response has been, I am not sure how 'I' can change the world's view on bigotry and prejudice, as it will take lots of energy and time from everyone before significant progress can be made. However, we can take immediate action by sponsoring and mentoring the next generation of professionals and leaders. They may not have as many professional scars and examples of psychological trauma as a result of systematic barriers that have both intentionally and unintentionally been put in place. I can't speak for them, but from my vantage point, the next generation has a much different perspective that is fuelled by hope and the belief that we can do better as a society.

So, what is the next step? How do we take action? Over the course of the last year, I've given some thought to these questions and my perspective is by no means a comprehensive solution, but it is a starting point towards **B**uilding and **R**edefining **I**nclusion through **D**evelopment, **G**rowth and **E**mpowerment – **BRIDGE**.

The pillars that I propose are five elements that support collaboration and advocacy in many forms. Each one contains actionable steps that positively impact our I&D journey. The blueprint is designed to promote progress, and it guides us from having tough conversations to realizing equality and representation. I share these elements because like with so many complex challenges, many times we don't know where to start. From individuals, community organizations or corporations, it is my intent to leverage this approach to anyone seeking out an opportunity to drive collaboration through diversity of thought.

1) **Connect Talent to the Pipeline –** Identify networks of potential candidates who are qualified or preparing to step into industry. Having a talent pipeline in place allows you to nurture and build relationships with prospective candidates. It's an intentional, proactive approach to identifying, qualifying and nurturing candidates towards securing career opportunities.
2) **Mentorships –** Mentoring is an active partnership between committed professionals to foster growth and career development. The aim is to establish a vibrant relationship that will either increase job and career satisfaction or prepare aspiring professionals to enter into new opportunities.
   - Twenty-five percent of employees who enrolled in a mentoring programme had a salary-grade change, compared with only 5% of workers who did not participate.
   - Mentees are promoted five times more often than those not in a mentoring programme.
   - Retention rates are higher for both mentees (22% more) and mentors (20% more) than for employees who did not participate in a mentoring program.
3) **Retaining Diverse Talent –** Retention must be intentional and leaders have an opportunity during every interaction to boost engagement through

valuing, challenging, rewarding and developing their diverse talent. I believe that this can be accomplished while also delivering some value creation to the bottom-line. If we can identify opportunities within our organizations and align our talent's strengths to those needs, then we can spark value-creating passion projects. These projects accelerate retention because think about it, who wouldn't want to use their sweet spot and skillset to solve costly challenges or resolve difficult problems.

4) **Empowerment through Engagement** – As unfortunate as George Floyd's murder was, it did ignite most of us to take some level of social action. But why do we need to be triggered? Can we take a proactive approach to social justice? Sure, we can all vote to support political action and policy reform, but we can also highlight the unsung heroes that are making a difference where they are. Let's promote examples of diverse leaders, professionals and students who are making an impact in their industries and communities.
5) **Outreach** – Let's build our tomorrow's workforce today with programmes and partners who support I&D initiatives. It starts as early as the elementary age and is reinforced throughout post-secondary education. We don't have to reinvent the wheel, as there are hundreds of non-profits, academic organizations and professional societies that need our help and input to accomplish their missions. We just need to get plugged into those groups, learn more about their visions, identify how we can support them and then volunteer our time to execute.

As a result of developing this blueprint of collaborating through diversity of thought, I founded a platform called Re.engineer, which is an inclusive community of STEM professionals that collaborate and share solutions. Our mission is to bridge the skills gap between professionals at all levels, and our professional development opportunities help our community members to achieve success in their careers. Through our growth, we have developed an interactive, digital magazine that helps professionals to see the possibilities within themselves. Even though the efforts to increase the amount of minority STEM professionals in industry have been outstanding, over the last decade, the number of minority students that actually graduate with STEM degrees have been declining year over year. Similar to the other innovative STEM platforms, our magazine aims to solve this problem by inspiring students to pursue STEM careers and by providing resources to those who have already started their journey.

In conclusion, collaboration can inspire innovation, enabling every person involved to bring their knowledge and skills to the forefront. Embracing different perspectives can be an excellent way to solve problems, improve efficiency and come up with new, exciting ideas that will ultimately lead to progress. All of the perspectives I've shared are tools of encouragement for everyone to think about how you will take action to create collaborative environments. This is just the beginning, but if we build a BRIDGE to accelerate collaboration through diversity of thought, then we can guide our way to (1) developing and retaining minority professionals, (2) securing the pipeline of diverse talent, and (3) creating value for our organizations all across the world.

## ABOUT THE AUTHOR

**Shadrach Stephens** is an Award-Winning Engineering Leader, Community Advocate and Founder of Re.engineer. He is a native of Baton Rouge, Louisiana, and a graduate of Southern University and A&M College where he earned a B.S. in Electrical Engineering.

Shadrach has progressed from several engineering roles for several Fortune 100 companies to where he is now the Site Reliability and Maintenance Director for the Celanese Corporation.

Outside of his corporate engagements, Shadrach started Re.engineer, a community that collaborates to prepare the next generation of STEM professionals. He has been married to his wife Marie Smith-Stephens for 16 years and they both enjoy volunteering and travelling with their son and daughter.

# 10 Five Unconventional Life Lessons for Anyone in Engineering

*Erin Gutsche*
Words with Purpose, Inc, Alberta, Canada

**CONTENTS**



*"What is possible and what is not is not your business. Nature will decide this. Your business is just to work for what truly matters to you."—Sadhguru.* I love this quote. It inspires me to dream big, and it reminds me to stay open to whatever happy surprises the universe has to offer. Too often, we impose unnecessary limitations on ourselves, and that's really the underlying theme of this chapter: don't limit yourself. Embrace who you are and be open to new possibilities.

## 10.1 A BIT ABOUT ME

Welcome, and thank you for joining me here! Let's get this fun rolling with an introduction.

Hi! My name is Erin Gutsche (pronounced *Goochay*). I'm a mechanical engineer licensed in the province of Alberta (Canada) and a Certified Maintenance & Reliability Professional. I'm also a technical writer, editor, entrepreneur, and – most importantly – the mother of two amazing little boys. I love music, laughter, dark chocolate, green tea, Formula 1 racing, the Rocky Mountains, and self-help books.

DOI: 10.1201/9781003336495-12

## 10.2 MY BACKGROUND

I grew up in a small town of less than 50 people in Central Alberta. I was raised by a loving family consisting of my parents, my sister, and a precocious cat.

My younger years were filled with books, puzzles, camping, and piano lessons. I was encouraged to pursue a variety of interests, and I'm grateful for all the opportunities I had as a child and as a teen.

As the oldest child by almost 5 years, I was called upon to help my parents with chores. I often helped my dad with what many would consider to be traditionally masculine tasks like stacking wood, building and assembling things, working in the yard, etc. I wasn't the kid who took apart toasters and televisions, but I was constantly asking my dad how different things worked and why. If there was a piece of furniture to be assembled, I was the first to grab the screwdriver.

My parents owned and operated a water well drilling and servicing company for over 30 years. As entrepreneurs, they ran the administrative side of the business out of our home office. The shop, which housed all the equipment and parts, was less than a mile away. From a young age, I was around drilling rigs, hoists, pumps, casing, and other mechanical parts. (Clamps and electrical tape made regular appearances in kitchen and office drawers.)

As I got older, I learned more about the intricacies of the business. I would help with things like bookkeeping, IT support, correspondence, and marketing.

I'll never forget the time I received an industrial version of an anatomy lesson from my dad. It was an inventory season, and I was counting different pipe fittings at the shop. Dad would call out the name of a part, and, with some assistance, I would find the appropriate pieces and start counting.

Some fittings were easy to recognize (e.g., a 90-degree elbow), while others were less obvious. At one point, my dad asked me to count the male threaded pipe nipples. I then made the mistake of asking how I was supposed to tell whether a fitting end was male or female. My dad handed me one of each and gave me a moment to compare the two. One look at the geometry and I quickly understood! Cue a hearty laugh from my dad and a red face for me!

Humor aside, I think it's fair to say that my career path was heavily influenced by my family and my environment when I was growing up, even if I didn't realize it at the time. To me, this is a good reminder that we won't always be consciously aware of how particular events, beliefs, and values can influence our futures.

## 10.3 ENGINEERS DO WHAT?

Engineering wasn't in my vocabulary as a child, and I was never one of those kids who knew exactly what they wanted to be when they grew up. In fact, the story of my decision to pursue engineering is rather lacklustre!

I remember sitting at the kitchen table in the spring of 2005 with a University of Calgary calendar (a book of all the degree programs and courses) laying in front of me. I would be graduating from high school in a few short months, and I hadn't applied for any post-secondary education yet. My mom, exasperated, kept pleading, "Just pick something!"

My high school marks were high enough that I could get into any program I wanted to. The problem was not knowing which program I wanted! I was intrigued by optometry but wasn't willing to commit to another 8+ years of education. My parents suggested engineering. They were familiar with the work of engineers, both through their business and through family members, and they thought it would be a great fit for my scientifically inclined mind.

Me? I wasn't sold. Even after reading the program description, I wasn't entirely sure what it meant to be an engineer. But at that point, I felt like I needed to choose something. So, I reluctantly agreed, and decided to enroll in the engineering program with the caveat that I could transfer out after my first year if I discovered that I really didn't like it.

Not exactly an inspiring start to a STEM career! But that's the beauty of life. It's perfectly imperfect. Our stories are all different, and that doesn't make them better or worse than any others.

In the next five sections, I'm going to share five more of the unconventional stories and lessons I've learned over the past three decades. Let's go!

## 10.4 LESSON 1: BE YOURSELF, UNAPOLOGETICALLY

I've spent most of my life feeling like an outsider. For decades, I felt like I was bits of everything and all of nothing.

I was a shy child who excelled academically. Ironically, my good grades were often a source of misery. Peers would give me a hard time if I received anything less than perfect marks, and they would celebrate if they received a higher grade than me.

My first experience with bullying was in the fifth grade. Other girls would exclude, ignore, and tease me, and say hurtful things for no apparent reason. My mom explained that these girls were insecure, and that their actions and words reflected how they felt about themselves. She was right, but those were difficult concepts for my 10-year-old self to fully understand.

By the time junior high rolled around, I found my place with a more inclusive and diverse bunch. What I loved most about these friends is that they always accepted me for who I was and never pressured me to be or try something I wasn't comfortable with. I could be an athlete, scholar, poet, musician, or some combination of all those things, and none of those roles fazed them.

In Grade 10, my circle of friends expanded to include a handful of young men I had known since kindergarten but hadn't spent much time with. They turned out to be an incredibly open and easygoing group! They accepted me for who I was with no labels or strings attached. They expanded my love of music, taught me the meaning of unconditional friendship, spent hours with me on MSN Messenger (on dial-up internet, no less!), and gave me more laughs and good memories than I can count.

In 2005, I left the safety and comfort of my friends and hometown to move to Calgary for university. (Talk about a culture shock: my first engineering lecture had six times more people than my high school graduation class.)

My identity took a hit. I went from being a poised high school valedictorian to a struggling 17-year-old, who lacked confidence and friends. I felt lost, and in a sense, I forgot who I was. (At one point, I was ready to drop out of the entire program.)

Thankfully, a fateful lab assignment turned that all around. It didn't take long for my lab partners and I to become great friends. They accepted me for who I was at my highest highs and lowest lows. We went to concerts, studied, and played hours of video games together. Through it all, I was always able to be myself.

While I've spoken a lot about being accepted for who I was by my peers, friends really aren't the key here. The heart of this lesson is that YOU need to accept who you are, first, foremost, and always.

I also want to mention that you might have a habit of feeling unaccepted if you were raised in a difficult home situation. I think this happens because (rightly or wrongly) we often form perceptions of ourselves through the mirror of others' reactions. But the good news is that habits can change. It might take some work and time to ferret out where those icky feelings sit, and why, but I have faith that you can do it if you really want to.

When you show up authentically, the right people will find you and vice versa. Let's face it – not everyone is going to agree with you or like you. There's nothing wrong with that! There's almost 8 billion people on the planet, and that's just a fact of life. Don't try to be someone or something for somebody else. Believe me, I've been there, and it isn't pretty. Anytime you suppress or hide part of who you are, you're limiting your experience. And it can be soul-crushing. (More on this in Lesson 3.)

Instead, just be you. All of you. The world needs you exactly as you are! Nerdy, quirky, athletic, tough, sensitive, strong, courageous, goofy. Doesn't matter. What matters is how you feel about yourself and how much you honor that.

There might be parts of you that you wish were different, and that's ok. You don't have to love your ears or your left pinky toe. There's a lot of power in simply accepting that which you cannot change.

You might be wondering, "How do I know if I'm honouring myself?" or "How do I know if I'm showing up authentically?" You'll know by how you feel. You'll find joy in the situations or people around you. And you won't have a weight on your shoulders, a knot in your stomach, or a niggling thought in your mind that something doesn't feel right. Instead, you'll feel light, confident, and open to the world around you.

One more important note: this isn't something that's going to happen overnight! Be patient and gentle with yourself. If you've spent much of your life trying to fit in or be someone other than who you really are, it's going to feel unnatural at first. (I'm 34 years old, and I'm still learning about myself and embracing who I am!)

As humans, we tend to label and categorize things. And while this can help us navigate the world around us, it's also very limiting. So, don't worry about the labels. You might be a girl or a woman in STEM, and that's great. But who are you really? Bring the energy, the emotion, and the ideas of who you are into the world. Share your light with those around you, and never apologize for being anything less than the magnificent human you are.

## 10.5 LESSON 2: ASK FOR HELP

Ask for help when you need it. It sounds so simple, right? But if you consider yourself to be independent, you might struggle with this. I know I did.

My Grade 12 calculus class sticks out as an example. It was an optional course taught by the high school math teacher. I got along well with him and enjoyed his teaching style, but for some reason, my 17-year-old self was determined not to ask for help with assignments.

I suspect there was some young pride at play. I also felt like I had a reputation to uphold as a top academic performer. How could I be the best if I was asking for help all the time?

What a foolish belief that was.

We were several weeks into the course before I finally made my way to my teacher's desk at the front of the room to ask for help. He could tell that I was reluctant to be there and put me at ease with a gentle joke about *finally* coming to see him. And that was it. Thirty seconds with him probably saved me 30 minutes of frustration and time wasted on one question.

This might sound like a simplistic and obvious example to some, but it was a big deal for me. My parents raised my sister and I to be independent young women. They wanted us to have the skills and the mind sets to go after whatever we wanted in life, without the need of anyone else.

Somewhere along the way, I developed a belief that asking for help was a sign of weakness. I've since come to my senses!

Asking for help isn't a sign of weakness – it's a sign of courage and wisdom.

It's about acknowledging your limitations and accepting that there's nothing wrong with those limitations. It takes courage to say, "I don't know," and "I need help," and "I'm not sure."

Funny enough, I ran into a similar situation in the university. I struggled with my first-year engineering statics course. Mix that in with painful shyness and my leftover pride from high school, and I was not about to admit my lack of understanding by asking a random stranger (i.e., classmate) for help. I was already outnumbered 10:1 by my male peers; I felt like I would be doing a disservice to the female gender by admitting my struggles.

There was no way around it; I needed to talk to my professor.

Ugh. Dread! I felt intimidated. Who was I to be asking an esteemed university professor for help? Shouldn't I be able to grasp these concepts easily and on my own? What was he going to think of me? Was he going to tell me I didn't belong in his class? Or in the university?

Thankfully, my fear of failing the course was stronger than the combination of my embarrassment and feelings of intimidation. When it finally came time to meet with my professor, I walked into his office expecting a haughty teacher who would sneer at the ridiculousness of my questions.

Once again, I was wrong. So very wrong.

The stoic professor, who was able to command a lecture hall of hundreds, turned out to be incredibly soft-spoken and non-judgmental. (There was another lesson for me: an innate ability to catastrophize any situation.) I received the help I needed that day, and from that point, I was able to approach other professors with much less trepidation.

By the time I reached the workforce, I was much more comfortable with asking for help.

I'll never forget my first day as a rotating equipment specialist for an infrastructure unit (i.e., water and wastewater treatment plant). I had only been out of university for 17 months, and here I was with "*specialist*" in my job description. My new job was to provide technical support to ensure the safe and reliable operation and maintenance of over 400 pieces of rotating equipment. I felt like a fraud ("Hello, Imposter Syndrome!"), but I was excited for the role and ready to learn.

This time, instead of falling into my old habits, I took a different approach.

I walked into the millwright task room to meet my new team, which consisted of four senior millwrights, each with roughly 30 years of experience. Let's put that into perspective – each of these technicians had been working for longer than I had been on this planet. And I was expected to provide *them* with technical direction! I knew that if we were going to be successful, we would truly need a team effort.

So, I walked in and said, "My title may be rotating equipment specialist, but you guys are the experts. I'm going to be relying heavily on your expertise to start with because I know I have a lot to learn. And if you're willing to teach me, you'll always have an eager student."

Much to my delight, the millwrights were warm and receptive to this. They took me under their proverbial wings and showed me the ropes. We became a fun little work family, and I'm forever grateful for the time they spent with me. I followed them into the shop and the operating unit like a shadow. Well, a shadow with tons of questions, that was!

It takes confidence to admit that you don't have all the answers. Don't let shame, pride, or fear get in the way of asking for help.

## 10.6 LESSON 3: FORGET THE WORD "SHOULD"

As a writer and editor, I love words. But "*should*" is one I could do without.

You see, *should* implies we are doing something wrong, either by our action (you *should* not do that) or our inaction (you *should* be doing this).

Here's the problem. These *shoulds* are usually based on someone else's rules, judgments, thoughts, and beliefs.

How many times have you been told that you should think, dress, look, feel, speak, or behave a certain way? How often have you been told this directly? How many times have you seen this in more insidious ways, say, through the news or social media?

When was the last time you stopped to reflect on your beliefs and values? Are they truly yours, or are they from your parents or friends? Are they a reflection of your religion, community, or education?

I believe success comes from authenticity.

When we suppress who we are to fit someone else's mold, two things happen:

1) We lose ourselves. We criticize ourselves and try to hide the essence of who we are. We might feel lost, stressed, depressed, anxious, upset, sad, or even angry.
2) The world loses out on what we have to offer. Our greatest capacity to create and innovate comes from feeling safe, loved, and at peace. This doesn't mean everything is all sunshine and rainbows. This means we have fully embraced who we are, and we don't have to spend precious energy or time hiding, worrying, or conforming. We can share ideas and thoughts because they're genuine and they come from our truest selves.

Pay attention to when you find yourself saying or thinking *should*. What happens if you replace it with something else? What is the basis of that *should*? Is it a fact, an opinion, or an untrue self-sabotaging thought?

Let's look at a couple quick examples.

- **I *should* eat more vegetables.** Do I want to eat more vegetables? Not really. I'm not sure I even like them. But, I know I haven't been feeling great lately. So, **I *want* to eat healthier**, so my body feels better.

  See the difference? The initial thought is uninspiring and critical. The final thought is empowering and uplifting. It shows a desire for positive change. Here's another example.
- **I *should* know how to solve this problem.** Well, part of me believes I should know how to solve this problem. After all, I have 4 years of university education. But who's to say that I should know how to solve it? Am I just being overly hard on myself? This is a new area for me. Hmm. **I *want* to do a good job** here, but I'm not sure how to proceed. I'm going to ask for support.

How we think has a big influence on how we act. Our thoughts drive our feelings, which drive our actions. But what if those thoughts are influenced by the ones we care about?

Sometimes the *shoulds* we hear come from the people who love us the most. They probably even come with the best of intentions. The problem is that nobody knows you as well as you do. So, listen to what your friends and loved ones have to say, but don't forget to listen to that voice inside you that knows how you really feel.

This lesson hit home for me when I had my first child. I quickly learned that parenting is one of those topics where everyone seems to have an opinion (and a strong one at that!). But I also learned that just because something works for someone else, doesn't mean it's right for me or my family.

Don't let *shoulds* stand in your way. If you find yourself thinking that you should or shouldn't do something, pause and ask yourself where that thought is coming from. Explore your resistance to it and understand what it is that you're really after.

## 10.7 LESSON 4: BE OPEN TO POSSIBILITIES

I'm not someone who typically likes surprises. I've always felt more secure with the world around me by maintaining a sense of control, whether real or perceived.

That line of thinking has led to some challenges and growth over the years.

My first year of engineering started in the fall of 2005. High oil prices meant the oil and gas industry in Alberta was booming. I was often told by well-meaning family and friends, "You'll never have a problem finding a job as an engineer!" (Remember that for later.)

First-year engineering at the University of Calgary follows a common core curriculum. At the end of your first year, you select your intended major and rank the remaining options. Chemical engineering was my first choice and not just because the oil and gas industry was at a peak. Chemistry was one of my favorite classes in high school, and I loved my first-year engineering thermodynamics and chemistry classes.

Well. You can imagine my disbelief and disappointment in the summer of 2006 when I read the letter that stated I was placed in the second-year *mechanical* engineering program. It turns out that chemical engineering was the most popular major that year, and my grade point average was 0.08 points shy of the cut-off. The letter kindly informed me that I would have the opportunity to complete my second year in mechanical engineering, then apply to transfer back into my desired program at a later date if I so wished.

I was sad. I was outraged. I was mad at myself. And "mechanical" became a dirty word in my vocabulary that summer. I felt as if I had no choice. I had persevered through my most difficult year to date, and I wasn't about to face the notion of quitting again.

So, second year rolls around and I proceed with mechanical engineering. And you know what? I liked it. I fell in love with fluid mechanics and was able to enjoy even more thermodynamics classes. Not only that, but one of my best friends from first year was also in the program.

By the time my second year ended, I was in a good place. I opted to stay in the mechanical engineering program and, funny enough, it turned out to be the perfect program for me. I took a turbomachinery design course in my fourth year and loved it. That course afforded me the opportunity to attend a gas turbine symposium with three classmates, and two of us were even awarded scholarships! I was also able to complete a petroleum engineering minor as part of my degree.

At the risk of sounding cliché, the moral of this story is that sometimes things work out exactly as they're supposed to, even if it's not what we originally expected or hoped for.

Oh, and that bit about never having to worry about finding a job as an engineer? Well, the oil and gas boom that coincided with the start of my university studies burst by the time they ended. I was able to land a role while still completing my fourth year, but many of my fellow graduates struggled to find a job after graduation.

The second example I want to share with you comes from my second maternity leave. I was fortunate enough to take an 18-month leave from work after I had my second child.

I loved being at home with my boys. But – and there is a *but* – I was starting to crave an intellectual challenge. I was also concerned that I wouldn't be able to return to my engineering role in a part-time position; I knew I wasn't ready to leave my boys for full-time work.

So, I decided to take an online course. It was an *Introduction to Editing* course offered by a Canadian university. What on earth prompted me to do that, you might be wondering?

Well, I've always enjoyed reading and writing. Archie Comics® and Nancy Drew® mysteries lined my bookshelves as a kid. (I loved making road trips to the Edmonton Costco® in those days. The store sold Nancy Drew® books as a three-pack, but I would often get in trouble for starting and finishing a book on the drive home!)

I also remember my mom encouraging me, at the tender age of eight, to write *shorter* short stories for my upcoming provincial academic achievement test. My practice stories were too detailed and took too long to write!

My love of reading and knack for writing continued into my teenage years. My parents recognized and encouraged my skills, and often asked for my advice when drafting letters and marketing material. Fast forward to life as a young mother, and I found myself contemplating a side hustle that would make use of my love for the written word. So, I started with that *Introduction to Editing* course while home with my two boys.

That one course turned into a second course, which eventually led to three more courses. And there I was, five courses-deep, home with two young children (who were one and three years old at the time), contemplating something crazy – starting a business.

I will be the first to admit that until that point, I never wanted to be an entrepreneur. My parents had asked on multiple occasions if I wanted to take over the family business someday, and my answer was always a hard no. I saw how hard they worked. I saw the ups and downs of market swings, the unpredictability of extreme weather events, and the challenges that came with clients and employees. I was not interested.

But all that changed when I had my boys. The world of small business had a new appeal: more time with my babies.

So, I went against the routine-loving core of who I had been and jumped in with both feet. On December 2, 2019, I stood at a local registry office and brought my editing and writing business into the world. I distinctly remember thinking *What the heck am I doing?* But, before I could change my mind, it was done. I paid the fee, received the paperwork, and suddenly, my little corporation was a real thing.

The next few months were spent developing my website, learning about marketing, joining LinkedIn, and redefining my value proposition. Oh, and did I mention I returned to my engineering role in an 80% capacity during that time?

Life has a funny way of bringing us not necessarily what we want, but what we need.

My business has grown from a labor of love to a profitable venture of self-expression and growth. It's combined my technical background with my love for the written word, and it's given me a sense of freedom and pride that I can only describe as uplifting. It's also resulted in me meeting so many amazing people, and that's really been the best part.

The moral of this one? Don't limit yourself to what you think might be right for you, or what you're most confident in. Be bold enough to take chances.

I like to think that the universe works for us, rather than conspiring against us. Looked at closed doors as opportunities to find something even better. And if your heart guides you in a surprising direction, listen to it. You might be pleasantly surprised by where it's leading you.

## 10.8 LESSON 5: PRACTICING SELF-CARE IS NOT SELFISH

As you may have noticed, I'm a parent, a recovering perfectionist, and someone who was raised to believe that hard work and self-sacrifice were the paths to success. If any of those resonate with you, I want you to pay close attention to this one.

Self-care is not selfish.

Repeat after me: self-care is not selfish.

We live in a hustle-and-bustle world. The internet and social media are at our fingertips. Data is shared in real time, work is happening remotely out of our living rooms, kitchens, and basements, and our achievements are more visible than ever before.

But there's a problem.

We think we need to earn rest. We lack balance. We turn to convenience and quick fixes to keep us going (think fast food, energy drinks, medications). We manage stress poorly. We get so used to being in a state of flight, fight, or freeze that we forget how to relax.

I'm as guilty of this as the next person.

In my first year of university, I was terrible at time management. I would underestimate how long it would take me to complete assignments, and I told myself that I worked best under pressure. I pulled one all-nighter that year, but nights with only three or four hours of sleep were common. The result? Burnout, stress, depression, and a desire to drop out of engineering. I didn't realize how much my lack of sleep was affecting my mood and ability to focus.

Things shifted as the year progressed, and over time, I became more aware of my need for proper sleep and nutrition. My second year was better (though far from ideal), and my third year was even better still. By the time internship rolled around between my third and fourth years of school, I realized how important breaks were.

While on internship, I worked a typical 8-hour day. With no homework, my evenings and weekends were free. I was able to spend time with friends, take up a new hobby (golf), and enjoyed life. By the time my fourth year came, I knew how important sleep was to my success and overall wellbeing. I procrastinated less, slept more, and studied more effectively.

With all that said, burnout made another appearance after the birth of my first child. My sweet little boy was a terrible sleeper; he didn't start sleeping through the night until he was 18 months old. Combine that with an inexplicable rash that covered his body and difficulties with food, and I was one tired, stressed mama.

I've since been blessed with another beautiful baby boy. At the time of this writing, my kids are 5 and 3 years old. Between balancing a nearly full-time engineering role, a small business, and two energetic kiddos, I've learned that I need to take care of myself. Sometimes that looks like a 30-minute workout. Sometimes it's a 15-minute meditation. Sometimes it's a few pieces of dark chocolate and a good book.

A counselor gave me a great analogy: the airplane oxygen mask. If you've ever flown before, you've probably heard an airline steward or stewardess explain that if the oxygen masks deploy from the ceiling, your job is to put your mask on first before you help anyone else (particularly children). You can't help anyone if you don't help yourself first.

The same goes with life. If you're tired, stressed, or burnt out, it's going to spill over and affect the people and situations around you. You might be short-tempered with your children or annoyed by your partner. You might struggle to focus on work or start to lose interest in the things that once mattered to you.

Now, this doesn't mean that you have to meditate for 90 minutes every day, go to yoga class three times per week, and only drink green smoothies. Not at all.

It means making time to rest and do the things in life that bring *you* joy. Don't do what other people think you should do – do what makes you smile. Do what warms your heart. And be honest with yourself!

It's okay to love your family and to want some time to yourself! It's okay if you're not the outdoorsy type and you'd prefer to spend an hour at the mall. And if you've been working so much and so hard that you don't even know what brings you joy anymore, that's okay too. Take some time to think about it and write down what comes to you.

Joy doesn't have to come from long, fancy vacations. It could be an hour reading a book. It could be a walk with your dog. It could be working on a vehicle. It could be an afternoon at a spa. Whatever helps you to slow down and reconnect with happiness and peace.

"But I don't have time for that!" you might be saying. Start small. Take 2 minutes between meetings to focus on your breath. Awareness is the first step to feeling better.

To me, health is multifaceted. I like to think of wellness as the sum of our physical, mental, emotional, and spiritual states. We need to take care of all of these states to be our best selves.

Maybe you're great at staying active, but you feel like you have an endless "to do" list that's weighing you down. Or maybe you're juggling a busy life but sacrificing your sleeping and eating habits in the process. Perhaps you're reassessing your values and belief systems.

Whatever you're going through, I also want you to remember that self-care isn't just you reading self-help books or talking to a friend. It also includes engaging professionals to help where needed, be that a psychologist, counsellor, massage therapist, physiotherapist, dietician, physician, you name it. There's no shame in taking breaks, setting boundaries, or asking for help.

The key is this: you can't be the best version of yourself without prioritizing yourself.

## 10.9 SUMMARY

Let's do a quick recap on what we've covered:

1. Be yourself, unapologetically
2. Ask for help
3. Forget the word "should"
4. Be open to possibilities
5. Practicing self-care is not selfish

I've learned these lessons the hard way through experience.

My hope is that you can keep these ideas in mind as you move through your life. Embrace who you are, don't be afraid to ask for help, honor yourself, stay open to the surprises in life, and prioritize your wellbeing.

Thank you for taking the time to be here.

I wish you all the best that life has to offer!

## ABOUT THE AUTHOR

**Erin Gutsche** is the founder and president of Words with Purpose Inc. She is a professional mechanical engineer in the province of Alberta, Canada, a Certified Maintenance & Reliability Professional (CMRP) with the Society for Maintenance & Reliability Professionals (SMRP), a Lean Six Sigma Green Belt, and a member of Editors Canada.

Erin can be reached by email at erin@wordswithpurpose.ca or via LinkedIn at linkedin.com/in/erin-gutsche.

# 11 STEMming across Three Continents

*Bralade Koroye-Emenanjo*
Fortune 100 Operations Leader, United States of America

**CONTENTS**



## 11.1 MY "RIGHTEOUS INDIGNATION": STANDING UP FOR MATH AND GIRLS

> I know many of you cannot derive $dy/dx$ from first principles… you are all here looking at me, you want to get degrees but you don't know math. You don't know algebra. I don't know how you will get your degrees. I challenge you to come down and solve it if you can…

The professor cackled and said, "Even the young men in this class are shaking, not to talk of a girl. No girl here can solve this. This is how I will wait and no one will solve this. Shame on you."

As he waited, he strode up and down the small stage, looking through the crowd before him, triumphant at the confirmation of his bleak assumption. This felt like a Goliath-taunting-Israel moment. I felt the hackles on the back of my neck stand and that was the first time I took a public stand for gender advocacy in STEM.

I stood up with a suddenness that was dizzying. I was halfway through the back of the lecture theater because I did not realize people rushed in to save seats hours before the actual class. I mean, it was my first semester at the university. I graduated from secondary school 6 months before my admission. At just 16, fresh-faced and chubby, I had sat somewhere at the back of the 1000-seater lecture auditorium at the University of Port Harcourt. To my young eyes, it must have seemed infinite, like 10,000 people could fit in it.

As I walked through the crowd, the lecturer seemed oblivious to me and was about to move on when the rest of the class started pointing me to him, almost like I was a signpost. He turned to look, startled however he quickly regained his composure. He

DOI: 10.1201/9781003336495-13

smirked and egged me on, patronizingly, "Come and try your hand", he said, his Nigerian English sounding heavier than it should.

I took the chalk, my breathing coming in short gasps; a mix of both indignation and nervousness at this very public pedestal I was on. I knew the same crowd who hailed me for a savior just seconds before would easily turn me into a laughingstock if I failed. But none of these move me as I started with the formula, $y = f(x)$. White chalk in hand, with those deft strokes, I wrote the 10 lines of algebra I had come to learn by practice and logic.

You see, math for me was just as fluid as English and grammar. It was a language I understood and that understood me. From my senior secondary school math classes, I had come to have a healthy respect and genuine interest in a subject that followed principle and resolved itself in every equation. I had come to know it as a set of logic that honored everyone who would follow its principles. And instead of resorting to cramming and memorizing things like my peers, I was a genuine student, allowing the numbers and variables to take me where the answers lay. Stating my assumptions and relying on principles as I progressed.

As I wrote on that board, watched by a professor whose name I did not yet know, I was transported to Mr Okrinya's math class in junior secondary school and the numerous classwork and homework he would have us do to become masters at mathematics. I remembered how he honored merit and principles. In his class, I did not know that I was a girl. All I knew was I had fallen in love; in love with numbers and letters, so deeply that and my conflicting love for both science/math and English/Literature led me to chemical engineering.

You see in Nigeria, my home country, typically one could not have As in science and math and then go on to study law or the humanities because it would be considered a waste. If you knew Math but not science, you went to business and accounting. If you knew science but were not so interested in Math, then you would study medicine, but if you were a whiz in both, then it was engineering for you. I am privileged to have parents who were liberal enough to think that this 16-year-old could make her decisions, but society had already conditioned me – I would be going into a career in STEM. I had considered medicine because of the prestige of the title, doctor; but one phone call to my doctor uncle helped me confirm that they actually learned with dead bodies aka cadavers, sometimes using a saw just to get to the brains. Needless to say, I was both grossed out and spooked. So, with Tipex, i.e., Whiteout in the US, I made the change to chemical engineering. The fact that it was a male dominated field did more to attract than deter me.

You want to know what happened that day?

Did I get my math right? You bet I did.

After I finished and handed the chalk back to the lecturer, he paused and went line by line peering at my logic and assumptions, like someone who had instigated a wrestling match and now would do anything to protect his ego after high-spirited chants and jeers at his opponent. I was somewhat amused that it took him that long to check my work. Then, he reluctantly stated, more like stuttering grudgingly, "You are actually correct…"

That hall erupted in the loudest cheer and at that moment, my advocacy work began. To think I would have remained in my seat if he did not malign all of the women folk. Since then, I have taken the matter of gender representation and the assumption of competence aka double bind personally. But even then, I don't think I realized how marked I was by the experience, the later years of my training and career would go on to say how much.

## 11.2 CULTURE SHOCK: OF TITLES AND NATIONS

From elementary school, I knew I would study abroad. When both your parents studied abroad, then that inadvertently becomes your highest aspiration, especially as a teenager and young adult. Grace and Seiyifa Koroye were the best graduating students from their respective departments at the University of Ibadan (think the equivalent of the foremost Ivy League institutions in West Africa at the time). The rewards for that achievement were government postgraduate scholarships up to PhD in the United Kingdom. It was the natural progression of first-generation graduates in a society that was beginning to value education and the trappings of success that it brought with it.

So upon graduation, as I completed the compulsory National Youth Service Scheme that Nigeria ran to make better citizens of its graduates, I started working on my plans. Even as I taught Chemistry and Math in the little village of Owena in western Nigeria, 500 km from my home in Port Harcourt, my mind contemplated shores beyond my continent. When I was not speaking to the largely rural population who still traded by barter in some cases, I spent my evenings evaluating schools and preparing applications for universities in the United Kingdom. I still remember how wide the eyes of my students got when I told them of the larger cities and universities. Several of them were so book-smart but had hardly been exposed to the world outside their county. I still wonder where many of my girls ended up, I hoped they did not get married off after secondary school. I hope they became the doctors and engineers that their wildest dreams portrayed. I may never know.

Upon ending my service year, I moved to the United Kingdom for postgraduate studies. I got accepted into the Queens University, Belfast, to study Process Engineering, a specialty in my major, Chemical Engineering. I was elated, walking on air as far as I was concerned. The euphoria I felt dousing the trepidation of being in a land unknown and an extremely unfamiliar system. I went from mastering the names of my professors from the university website that I had studied in detail to actually meeting them in person. That was where I received my first culture shock.

"Call me David," Dr Roberts said, stretching out his pale, well groomed hands to me. Between wondering how a chemical engineer's hands were still so curiously supple and contemplating how to respond, I remained momentarily frozen.

The man before me was a highly accomplished researcher, leading a globally acclaimed research center, specializing in complex chemical reactions. He did not just have a PhD; he was a distinguished professor. Why would he want to downplay his achievements to a girl of no consequence like me? I had seen them do so in movies, but it was still surreal in the moment.

In Nigeria, titles were of significance, especially those earned through rigorous academic study. Beyond the conventional Dr and Professor, some Pharmacists and Architects would add appendages like Pharm. and Arch in place of the traditional Mr and Mrs/Ms. that is the extent of my meritorious culture. It would have been egregious to address any of my lecturers at the University of Port Harcourt by anything other than their title and last name. Those who did not yet have their PhD got a respectful Engr. Otherwise, a simple Sir or Ma would suffice. Formal was proper in the culture I grew up in.

I learned though, and in the course of my career beyond Belfast, not to underestimate anyone who introduces themselves by their first names. For all I know, a PhD may be lurking in the repertoire of their accomplishments. If I had a PhD, I wonder where I would lay, somewhere in the middle… maybe Dr Bralade. I still smile as I think of my introduction into the informality of the Western workplace and academia. Different strokes for different folks, names all valid yet interestingly strange to culture transplants.

## 11.3 CULTURE SHOCK: A DIFFERENT WAY OF LEARNING

As the only black girl in my class, especially from Africa, it was difficult to be accepted into the cliques in my master's program. Everyone was extremely polite, but I knew I was not really seen or engaged. Try as I may, my naturally extroverted-self did not seem to break through as I usually would so I found myself working more on my own. I felt I would have learned the system better if I had been more assimilated. It seemed okay at first but by my final design project, it became apparent that my learning style was different and could pose a problem I had not anticipated.

During my Bachelors in the mid-2000s, there were no computers in my department for students. So, all the calculations I used for my schoolwork, in thermodynamics and mass transfer, were by hand. Therefore, when I was assigned the heat exchanger design in my design group project, I fully assumed I would do the same. I proceeded to start the tedious calculations, some resulting in as much as 20 iterations to determine the equipment specifications. I did this tedious work with the sense of fulfillment that comes with this brain numbing repetition.

When I presented my results to the professor, he questioned the validity of my work because it was manual. He could not believe I did not use the HTRI software (Heat Transfer Research Inc) and wondered why. I felt he could have at least inquired about why I went that route, given that I was on a distinction GPA. Instead, I felt guilty until proven innocent. I have since come to learn about the prove-it-again and performance biases (see Lean In) faced by women and people of color; and it depicted this scenario for me. Now, they are easier to recognize.

Thankfully, I had the gall to again advocate for myself. I proved the accuracy of the result based on formulas in textbooks. We were both in alien worlds, he was incredulous that I would go manual, and me defending my precise, double-checked calculations. Eventually, he let it go, seeing that my formulas were correct. However, I wonder why he did not bother to recalculate it all by hand as a thorough Nigerian professor would have.

Looking back now that I have mastered the HTRI tool and used it for several years as a process engineer, I chuckle at how much time I would have saved myself. To put it in context, while my rigorous calculations took days, a HTRI initial result would have taken about an hour. Also, I could change any variable and not have to do it all over again. Talk about the convenience of technology.

I got my first glimpse of implicit bias and how it manifests beyond the consciousness of the mind. It does not make it easier to bear if you are on the receiving end. It is humbling to know that we all have them and as we do the work of uncovering them, STEM fields particularly will have a more inclusive culture for women of all ethnicities.

## 11.4 CAREER LEARNINGS: THE MORE THINGS CHANGE, THE MORE THEY STAY THE SAME

Moving to the states after my master's degree was a dream come true. I was marrying my college sweetheart who had immigrated to America in our second year of university. Home was where he was, and the American Dream beckoned. I settled into the diverse and vibrant Houston, landing a job at a chemical manufacturing company. There, I got introduced to employee resource groups, what I now believe is the people's lifeblood of any company that is serious about inclusion.

At one of my first events with the women's ERG, I got introduced to the Harvard Implicit Association Test that evaluates unconscious bias in a person. It was interesting to me. From all my years watching American Television and seeing the seeming equality on screen, my naivete led me to believe women in the US were just as equal as men. So, it was a shock to me when over 80% of the program participants tested favoring men as STEM career professionals vs. women. I tested as having no preference for either gender. My group was amazed especially given the stereotype of the African patriarchal society. Then, I realized that growing up to meet both my parents with advanced degrees and lecturing at colleges of higher learning, helped me adopt the idea that men and women are equal.

Even though I later got to meet the sometimes-blatant disregard for women, my foundation was set. What amazed me that day at work was the possibility that this problem could be global. I had gone through three continents and experienced the pervading inequalities across cultures. The more things change, the more things stay the same.

As my career has progressed, I have personally faced even more hurdles, like when my leader at a time gave me a low performance rating because he said, "you should be grateful considering you worked less time than your peers and had two children in consecutive years." He actually encouraged me to be "fair" to my colleagues. I knew this to be against company policies and I took him to task with Human Resources. Ultimately, it got resolved, but feeling undervalued was something I dealt with for a while then.

I hope more people are beginning to appreciate that when a woman takes a maternity leave, it is not a vacation, and neither is it a "time off." This is arguably the most difficult job for any new mom. To come back to the workplace and face maternal bias is unacceptable and we, as a society and an industry, can do better. Since then, I have gotten involved in advocating for and changing policy for parental leave in the workplace and in America. In 2018, I continued that advocacy in the halls of congress to

US senators and congressmen under the auspices of the Society of Women Engineers on their Capitol Hill Day.

After a decade in the industry in a variety of roles, I am most active in removing the obstacles in women's paths so they are not blindsided by what another woman would have gleaned in her career. For example, I never knew I needed a mentor or how self-promotion worked in the American workplace. "Keep your head down and do great work. Your leader will notice," was my cultural context for performance appraisals. With my learnings, I have mentored, championed, and volunteered for this cause.

I also started seeing the lack of diversity in some of our women's groups in STEM. The phrase "women and African Americans" became problematic for me because it obliterated a whole group of people from the diversity and inclusion conversation. Minorities that have intersectionality at gender and ethnicity face biases from both ends. Their representation was not included in the metrics companies judged themselves by. I began that crusade at my company and in women in STEM organizations – passionately advocating for representation in leadership, in programming and in recruiting. Sometimes, it makes people uncomfortable, but the pursuit of equality is worth a few squirms. It is the price of progress.

From my affinity to Math to my advocacy for women, the course of this Nigerian woman's journey in STEM has been fraught with culture shocks across three continents; successful with a rich career progression yet grounded in the sense of duty that a legacy of leadership demands. I have still managed to balance the love I have for numbers and words that I grew up with. Who knew a thing like podcasts would exist decades ago? Now I champion African women in STEM on my platform, STEMafrique podcast using stories to pique interest and data as my basis for advocacy. The more things change, the more they stay the same. The journey continues.

For every female scientist, engineer, technician, mathematician reading this, know that your story is valid and no, you did not imagine those things and yes, there are many rooting for you. Let's send down the elevator for the next generation. There is strength in numbers. We make a difference. Share your story today.

## ABOUT THE AUTHOR

**Bralade** has studied and worked in STEM fields across three continents in a diversity of work and ethnic cultures. She is currently completing a PhD in Leadership Studies at the Our Lady of The Lake University and is a Fortune 100 Leader. She holds degrees from the University of Port Harcourt and Queens University Belfast in Chemical and Process engineering. She is currently a Fortune 100 Operations Leader. A gender, culture, and STEM advocate, she led the Society of Women Engineer's African American Affinity Network and is the chair of her company's women's employee resource group. She is a podcast host, speaker, author, and coach. Her podcast, STEMafrique, is produced by Ebony Podcast Network. "A Line in The Sand," a bestselling short story collection, is her first book.

# 12 The Chemistry behind Applying STEM to Horses and Cage Fighting

*Jade Thompson*
PCMS Engineering, London, United Kingdom

**CONTENTS**



Science – Technology – Engineering – Art – Maths

Your whole life is supported and sculpted by STEAM. From the clothes and makeup you wear to the music you listen to. The perception of working in STEAM is often misconstrued, as is the journey from school to working in STEAM. It's not all mechanics and computer science.

Education gives you the basic requirements to start your career in STEAM, which is where the real learning begins. Let's say the STEAM industry is like a group of books; each specific industry is just the title, then you have the chapters to choose from, you pick a character and, who knows, you may even create your own?

My career has expanded in so many ways over the last 10 years. I have discovered so much, however, the biggest learning curve was realising just how much I can utilise STEAM in my hobbies. I am a passionate and a competitive horse rider. Utilising STEAM has allowed me to understand and monitor the biomechanics of my riding and my horse's movement. This helps me to understand our partnership in a whole new way and analyse how we can improve based on evidence and data.

DOI: 10.1201/9781003336495-14

## 12.1 INTRODUCTION

I'm going to start with where I am today.

My name is Jade Thompson, and I am currently a Lubrication Development Manager for a condition monitoring company called PCMS Engineering, which is a part of the PCMS Eng Group. My position encompasses a lot of roles, mainly technical support for the whole company and our clients. As an approved trainer, I have the pleasure of sharing my knowledge with clients and co-workers. I help with the management, development and growth of the Lubrication and Oil Analysis side of the business. I seem to wear many hats, which all fit in their own way as I grow and develop my career. Enough about my work life, my personal life is much more exciting; I have three horses, crochet from time to time, I used to compete regularly in boxing and mixed martial arts (however this is now more of a hobby!) Of course, my partner must be thanked for his support with not only my hobbies, but my career, as this tends to engulf me from time to time during large projects!

## 12.2 STEAM-DUCATION!

How did I get to where I am today?

The education system in the UK is keen to push students into STEAM subjects, they try hard to enable students to understand the careers associated with STEAM. However, if you had asked me at 14 if I would be working with oil and grease in an engineering setting? I would have responded with 'absolutely not!'.

I loved all three sciences as well as drama, history and physical education at the comprehensive school. Looking back, I had no idea how much they would mould my future.

Drama and Art are seen to be subjects, which only a minority can pull success from; however, they provided me with a set of skills which allowed me to develop key skills required through my career such as public speaking, creativity, training, marketing, management, design and much more!

Science was my go-to subject, especially biology, which gave me the thirst of investigation, methodical approaches and conclusive ends to projects. I have utilised science so much through my life not only in my career, but also my personal life. So much so, certain aspects I am only just realising as I write this chapter.

Physical education went hand in hand with my active personality and love for biology, teaching me the importance of a strong team and leader, all while learning about the biomechanics and physiology of my body.

At school, I was not thinking about long term careers and what was coming next, I was very much a 'live in the moment' child! I was hard working and ambitious; however, I had no idea what I was working towards.

A-levels came next. I was fascinated with Biology so, I chose this subject along with Drama, IT and Physical education at A-level. What a challenge A-levels were! Surprisingly, biology yet again, came out on top as my favourite subject! I really do believe the huge step from GCSE to A-level really helps students understand the importance of loving what you learn, which also helps mould what we do next. During my A-Levels I started working for a local pizza shop, which gave me my first insight into the rush of fast paced work. I continued to work for the same takeaway restaurant until I finished university 5 years later. During this time, I grew from kitchen staff to takeaway supervisor. STEAM helped me through without me even knowing it. IT, ART, Science and Drama also allowed me to help the shop owner be innovative with new dough mixing methods, optimised cooking times and developing new marketing material.

So, University application time came, and my world seemed to fall apart, I had no idea what I wanted to make of my life. I could have done with a mentor in the STEAM industry back then, or even better, this book! I went with my gut and found a degree associated with the golden subject of Biology. To top it off, I stumbled across an Equine Science degree! I was suddenly back on top of the world with my love for horses and passion for science all wrapped up with an Honours degree on top!

## 12.3 SADDLING UP WITH STEAM AT UNI!

During my degree was when I started consciously making connections between STEAM and my personal life and goals. I merged what I was learning of biomechanics with my riding career, giving me and my horses the best chance of success and harmony. Laboratory sessions were daily at the university, especially during my dissertation and I felt at home! The chemistry, physics and biology all rolled into one, with a strong focus on equine. I took to the laboratory work and decided to carry out a scientific experiment on parasites for my dissertation. This led me to organise my career path with the ambition of completing an MSc in the equine parasitology field.

## 12.4 STEAMING UP AT WORK

I needed lab experience for my MSc, so I hunted for a laboratory job close to home. This was around the time I started competing with my boxing and MMA. I found a job at an oil blending plant as a Laboratory Technician. This was a whole new scientific subject which I had no idea existed and I challenged myself to learn everything I could. Who knew oil was so interesting and complex? I worked at the blending plant for 5 years, and within that time, I developed my skills into covering roles for numerous jobs within the company. There was no permanent progression for me at the oil blending plant, so I decided to move on to my current employer PCMS Engineering. I am now in my sixth year of employment here and do not regret sitting my MSC. I have loved every minute of my career with both companies. PCMS Engineering gave me another opportunity to learn a whole new side to lubrication in its used state and this is where engineering came into my life. To understand the performance of the lubrication, we must understand the performance of the equipment it services. I am still learning every day and I have been lucky enough to gain many accredited qualifications, including other condition monitoring technologies such as Vibration Analysis. My career path through PCMS Engineering has allowed me to develop skills, not only within STEAM, but with management, marketing, training, health and safety and much more. From a laboratory technician to a technical lead for the company, it's been a huge learning curve with a new scientific subject and new responsibilities and, ultimately, balancing my life goals.

## 12.5 FIGHTING WITH STEAM

My fighting career lasted around 6 years. I do hope to one day fight again; however, for now, it is a hobby! Fighting gave opportunity for more conscious uses of my STEAM education: eating, technical training, conditioning all parts of the process. Understanding the bone adaptation process allowed me to tailor my rest periods with the type of strains I had been putting on my body during conditioning work. I analysed my technical training by adapting my biomechanics system to work for me and my horses. This also allowed me to ensure I could monitor any changes in my body from fight fitness and how this may affect my horses or my performance during riding. Eating the correct foods at correct times of my fight campaigns meant I could enter fights at the correct weight with the most energy possible.

## 12.6 SADDLING UP WITH STEAM AT WORK!

I know what you're thinking ... What has oil condition monitoring got to do with horses? My current job role is more about asset reliability as a whole. There are lots of technologies and tools, which help us monitor and diagnose asset health. Thermography being the main technology I can take home with me. I can use thermography to assess my horses' health and recovery progress after an intense training programme. I can monitor how his legs and hooves deal with new surfaces. How warm he is in the winter, after riding. I can assess whether I need to alter any of the variables. The most vital help from Thermography has been ensuring my horses' saddles fit correctly and establishing my weight distribution while riding. Am I putting too much weight down their right side? Will this cause them discomfort? If so, the biomechanics jump in to help! One of my mares in particular suffers from a disorder called Laminitis which causes the inner hoof to change and can lead to permanent skeletal change and lots of pain. Laminitis is hard to diagnose until the visual symptoms have kicked in; lameness, hooves hot to touch, horse not willing to move. I have managed to utilise Thermography to track when the temperature in her hooves starts to rise way before I can feel or see a difference. I can then change her routine and eating habits to ensure her condition doesn't get any worse and keep her in a pain free comfortable existence, which of course makes me very happy.

My take home message from this chapter would be to follow the subjects you enjoy. Take more time to think about where and how the daily things we take for granted have come to be. For example, who invented my perfume? Where was it made? How was it made? Who looks after the equipment that made my perfume?

## ABOUT THE AUTHOR

**Jade** is a Lubrication Manager for PCMS Engineering and an approved trainer for the British Institute of Non-Destructive Testing. With a dedication to improving things, Jade's career path naturally went in the direction of the reliability industry. From a science loving school child to a member of senior management, she has managed to transfer her skills and knowledge between her work and personal life.

# 13 Finding Your Way

*Liana Roopnarine*
GLEAC, Abu Dhabi, United Arab Emirates

**CONTENTS**



From a young age, I contemplated, 'What can I offer? How can I make society a better place? What is my true purpose for spending time here on this earth?' More frequently than normal, the answers to these questions are recognized as a part of destiny, as an outcome of self-realization, as putting pieces of the puzzle together. For me, it's constant evolution, multi-dimensional, built with imagination and even regenerative. I like to consider it, rebirth. The distinctive feeling of always wanting to give back to the society, to contribute to economic growth and worldly development apart from my passion for mathematics, unlocked my path and guided my heart towards pursuing a career in STEM. I always imagined being different, unique; in fact, I adore when people call me weird, and I never liked being part of the 'in-crowd'. Simultaneously, conversations with my experienced brother-in-law helped me put the story displayed deep within my brain, into black and white right where it needed to be to make it clearer, more concise and actionable. Fast forward to 2022 with over a decade in the oil and gas energy industry and having now transitioned to an AI tech start-up, I feel not one ounce of regret but only feelings of fulfilment, contentment, quest, resilience and perseverance.

## 13.1 LET YOUR VOICE BE HEARD

Pulling into the parking lot before 5:30 a.m. daily was like the icing on the cake, you beat exorbitant traffic jams and get to choose which 'first come, first serve' parking spot you wanted. This was an apparent luxury, if I may because the reserved spots were for line managers and those at the top of the organizational hierarchy. Most importantly, I enjoyed the unruffled peace and office stillness at that early hour. It allowed the best digestion off my cup of tea with a sweet momentary daydream at my desk before I dashed over to get my daily movement in, either at the gym next door or complete a joyful jog around the world famous, Queen's Park Savannah (QPS) in

DOI: 10.1201/9781003336495-15

the beautiful capital of Trinidad and Tobago, Port-of-Spain. I must admit, daily exercise for me is a top priority, it sets my mood, energy level and powerful momentum for the best day, each day. I then return to the office, devour my home-made hearty breakfast bite in front my desktop and set-up for a sustainably productive day.

As a junior engineer without a structured graduate training programme, you find yourself visiting managers' offices pleading to be noticed, probing to be challenged because your brain is like a sponge waiting to become saturated to take you on a learning journey you only dared to imagine. At my first corporate job, straight out of university, the first few months were filled with boredom and me questioning myself, 'Is this really what it's like to work in this majestic glass-cladded building? How do you even get to a leadership role with tasks like these?' I mean, my parents instilled in me from a young age, the concept of dreaming, dreaming hard, with deep determination and to dream BIG. I was given simple, repetitive, administrative tasks. However, it didn't stop me from seeking more and letting my voice be heard.

Six months into this junior engineer role, I was informed by management as a result of the 2008–2009 economic downturn, and reduced revenue from declining oil prices, I would be retrenched under the basis, 'last-in, first-out'! I was appalled, disappointed and hurt. My heart sank. In my brain, I could not understand, 'How could they do this to the lowest paid employee with no extra perks? What positive, economic impact could my removal bring to the company?' Well, I made up my mind, I wasn't letting them off this easily, honestly, it made no sense to me. Don't you agree?

During the one-on-one meeting amongst my manager, the HR manager and myself, I let my voice be heard. I listed numerous initiatives that could actually have real economic impact other than my miniscule salary. They let me speak, I was passionate about my ideas. I even suggested retaining me on the team so I can at least gain 1-year industry experience and I will depart. They looked at me in awe, the HR manager thanked me and I left the office. The retrenchment letter never came. I was the only one from the company, who survived that retrenchment out of approximately 15 people.

I used my voice to change a perspective and won. *Communicating and negotiating effectively* can be the game changer for you. Giving yourself the space and time to plan how you get your story and or ideas across is critical to moving you forward, one step at a time. I encourage you to use your voice, you will never know unless you ask … go for it.

## 13.2 YOU DON'T EVER NEED TO FIT IN

Walking through the corridors, feeling eyes on you, some whispering, others blatantly staring you in the face. Wonderous looks off, 'Why are you still here?' Can you imagine what that must feel like for a junior, female engineer with no role model to look-up too? I stayed focused and thought it was better to ignore these and get to where I wanted to be. I paid attention to my inner voice and expelled any negative energies in my external environment. One year went by and I was still employed. It felt truly amazing. I had beat the odds and was now referred to a project team who were recruiting team members for one of the largest projects in the company. My determination, creativity and persistence opened these doors for me, something which I hold dear to my heart.

My working environment improved drastically. It was more intriguing and facilitated an enhanced space for a young professional's growth and development. I was given a small project to showcase my capabilities and competencies to the project team responsible for the largest retro-fit offshore brownfield project at the company. The task at hand was exciting and gave me the opportunity for a one-on-one session with a key fundamental sponsorship, which could advance my career. I was not letting this go. I gave it my all in preparing for the presentation and during the presentation delivery. After that session, I was selected as a core team member for the project. A small win, but for me it was like winning the jackpot. I finally felt noticed and valued for the person I am and the positive impact I can impart on the business.

Being a core team member on a four-year project, the largest retro-fit offshore project in the company was the best thing that could happen for me at that point in my career. I was able to work and collaborate with all departments within the company and industry as whole. This allowed me not only the opportunity to understand fully the operations of the business but also to contribute my ideas. Our team met with all the large international engineering firms. I got access to relevant training and development to upskill and use my potential to the fullest. After all, working on the largest offshore project in the company made you a celebrity, well almost.

Here, I want to share something from deep within, I believe, this opening materialized mainly because of my initial experiences on the job with the lack of support from leaders and a safe space to voice opinions. The middle management team did not understand who they were as individuals or even as a team. They were unable to imagine for themselves how to make full use of talent. Perhaps, they were totally engrossed in themselves and their own progressions. Hence, they were blinded by this.

Moreover, communication, core judgement and decision-making skills were strikingly lacking. For example, I would often walk into a common area and feel as though I was not welcomed. Senior professionals made comparisons and formed classifications based on the brand of mobile phone you possessed, which high school you attended and even which engineering degree you completed. Yes, somewhat unimaginable, but true, we are faced with these biases daily. You repeat to yourself 'not to take it personally' and move on since the behaviours, unfortunately, were a true reflection of those individuals' life-long experiences.

In this sense, it is critical that organizations train their staff and leadership team on the importance of building diverse, equitable and inclusive teams. Research has shown the benefits of promoting inclusivity and the maximum return on investment it can impact on business growth. If you're not sure about this, then Google it. Ultimately, not fitting in gave me the space and time to find my true self, build my self-awareness and guess what, it totally worked in my favour.

## 13.3 BE YOUR AUTHENTIC SELF, THE STARS LEAD THE WAY

When I commit to something I am passionate about, I ultimately give my best. Being on the core team of the largest offshore project in the company valued over 30MUSD (for some this may be small but for a field producing under 20K barrels of oil per day [BOPD] this is big), it's your time to shine. So, I put on my dancing shoes and gave it my all. This was a flagship EPC (Engineering, Procurement and Construction) project and I gave it my all from conceptualization straight to commissioning and start-up of the plant.

This project enhanced my critical thinking skills and embedded deep within my core the human skills required to become a better leader. Brainstorming, reviewing and approving engineering drawings, interpreting and analysing data, solving problems and envisioning them come to life right before my eyes, brought pure contentment. Something which particularly stood-out to me was building a global, collaborative network of people-centric relationships who eventually became your friends. From Port-of-Spain, Trinidad and Tobago to Tulsa, to Houston, conducting and leading international Factory Acceptance Tests (FATs) on equipment skids and witnessing the full installation and operation offshore, was outstanding.

The relationships which were forged from start to completion of this project are those for a lifetime. I learnt from the start how to communicate effectively with all stakeholders. It is key to ensure that everyone feels connected to the main goal from initiation. Moreover, fully understanding how each individual contributor fits and makes the puzzle complete is of paramount importance. Daily meetings, keeping abreast of all activities while creating a safe space for sharing of ideas, voicing opinions, allowing constructive criticism and simply being heard, took us closer to the end goal one step at a time. Noticeably, team members would come directly to me to ask pertinent questions, to request relevant information to support taking their task to completion. As a young female, my individualistic leadership style created this safe, adaptable space, which ultimately led to the successful completion of the project and numerous growth opportunities for the business.

At the company's first energy conference in Spain, I represented the Trinidad and Tobago team and presented on the project. It went extremely well, and I was finally able to fight my fear of public speaking. Oh, what a confidence booster. It's so important to hold that fear, stare at it dead-centre and knock-it out of the way. After my presentation, one manager at headquarters made it his business to walk-up to me, shook my hands and congratulated me. An energizing connection sparked, we built our professional network and supported each other and the company well into the future.

Taking on responsibility, staying committed, holistically leading and learning every step of the way did not come easy. I do believe, being my true self, made a difference. My manager, my male sponsor, my ally, saw it, he bet on me and gave me the opportunity to prove myself. And so, I did. My excellent performance was rewarded annually with numerous salary increases, recognition prizes and visibility. Upon building my credibility, my name was being announced at the headquarters in Spain. I screamed with happiness!

## 13.4 YOU WILL NEVER KNOW UNLESS YOU ASK, SPEAK UP

During every performance review session with my manager, I was determined and consistent to inquire about ways that I could improve and bring more value to our department and the company. It was an open, 360-degree conversation; specifically, with this manager, I felt safe, valued and appreciated. This is fundamental to leadership and bringing out the best in any team. I allowed my manager to speak, and my manager did the same. While we had these annual reviews for four consecutive years,

I would reiterate a significant point which I held close to my heart. The aspiration of one day working at our headquarters or at another subsidiary.

This is where I am living proof of setting aspirational long-term goals and seeing them unfold right before your eyes. I asked for it, the universe aligned and magic happened. Upon successful completion of the project, which contributed to my foundation as an engineer, leading in the offshore oil and gas industry, I was expatriated to our headquarters in Spain.

This was a dream come true for me. I truly love experiencing and diving-deep into new cultures. Living in an unfamiliar country with fresh perspectives is soothing for my brain and contributes to my cognitive development. This is my opinion, an open heart and an open mind. Unfortunately, for some, there was bickering about linking my age and expatriation but that did not get in the way, and I've put that behind me. What is made for you can never be taken away from you. Seek out and the answers come your way, perhaps not exactly when you expect it but when you least expect it.

I spent approximately 2 years as part of a diversified global team within the Business Development department at our headquarters in Spain. I must admit, the entire experience was intensively intriguing. I worked closely with subsurface discipline engineers, senior engineers and project managers to develop oil and gas surface facilities designs that would promote business growth and key value chain optimization. State-of-the-art software and data analytics decision making were some of the tools implemented. My learning journey was adventurous each day and made the expatriation experience fulfilling.

The environment at headquarters was very nurturing, encouraging and superior. A colleague and I worked together to improve an internal process, which we were able to fully automate and generate more accurate and precise analyses. We undertook this initiative upon ourselves and received high praise from management for our forward thinking skills. In addition, the environment offered me the opportunity to complete my Master of Science thesis from the University of the West Indies, remotely even before the pandemic, deepening my learning agility. After all, learning is a timeless, enjoyable journey.

I am forever grateful for the opportunity to live and work in Europe. It taught me the true essence of valuing time and establishing an expedient work-life-balance. We all get these 'ah-ha' moments in life, which help us pivot. I encourage you to reflect and identify specific turning points in your life. Specifically for me, it not only strengthened my collaborative and leadership skills but also developed my self-direction skills. It reinforced the saying 'work hard, play harder'. The work-week was mainly focused on work activities, while during the weekend, I got myself captivated in some form of exploration activity. My weekends consisted of taking a road-trip to another town, hopping on a plane to another European city or simply enjoying the local parks, concerts and museums. This experience will forever live on within me and has modelled me to be the strong, independent, focused, innovative leader I am today.

## 13.5 MAKE YOURSELF VISIBLE

It was the year 2016 and yet, another downturn in the oil and gas industry. This meant my expatriation in Spain was interrupted. I was disappointed and heart broken.

Fortunately, life has taught me to never let disappointment get in the way. I learnt from my real-life experiences, the importance of agility, flexibility and resilience, what better way? I took the news and began formulating my rebound for my return to my home-base in Trinidad and Tobago.

I requested meetings with Human Resources at headquarters and with my home country, Trinidad and Tobago in advance. I was waiting on no one to determine my future. Clearly, the resultant empowerment from self-direction, emotional IQ and mindfulness. The company was offering me a return to my 2-year-old position but with tedious, unchallenging, unexciting projects. Honestly, I did not want this.

I contemplated and brainstormed further and requested for all the possible openings I could fill. I requested to move to the planning and strategy department but got turned down. Staying focused and looking straight ahead to conquer my goal, nothing would stop me. Eventually, something opened in supply chain management (SCM) with my first female manager. I was impressed and grabbed the opportunity.

Feverishly taking on this administrative role in SCM, I must admit, complemented my sound technical background. I learnt tons and was able to contribute my ideas and implement process changes. This owed ultimately to the fact that I came from an engineering and operations background and understood the entire business model end-to-end. As such, this contributed to an improved decision-making process and the ability to set-up new contractual agreements reflective of the companies' new business model in an optimized time period. In this role, I was able to refine my negotiation skills and creativity. I was super proud of my contributions and the evolution of the relationships that I led between the company, contractors and the supply chain industry.

I spent approximately 18 exciting months in this role and requested a return to engineering and operations. What led to this? My unique desire and hunger to learn, to continuously re-invent, to seek out new exciting opportunities to build myself. In the SCM role, I reached the peak of my learning curve. I was now, simply taking control of what I was in control of. Makes a lot of sense, no?! Well, I am living proof of it and I encourage you to enforce practicing this. Out with the old and in with the new. I was expeditiously reassigned to the operations department as a Project Engineer under the new operations model. We also had new owners, new managers, thus new leadership. This got me truly excited, it fuelled-up that inner strength, that power, that fire.

I must admit, holding this new position was one of the most impactful professional experiences I gained in my entire career. Here's why, I simultaneously led and managed over five (5) re-vamp offshore projects and continued to take on more responsibilities as projects were successfully completed. These victories were attributed to comprehensive strategy development, project planning, creation of SMART goals, key stakeholder management and boosting organizational culture. Seems all too common and obvious right? Seriously though, it worked with the right direction from upper management and my very own unique leadership style.

## 13.6 INTROSPECTION

It is amazing to smash goals! It boosts your confidence, highlights you as an experienced expert in your industry and grants you the opportunity to mentor. I spent approximately 20 months out of my 11 years in the Oil and Gas industry in this senior project management role and felt like it was time for me to move on ... move on completely. No more creatively requesting to shift departments and roles. I took the decision to gracefully exit the company and focus on a higher purpose in my life. It just felt right.

Self-reflection and inward introspection generated this moment for me, and I am so grateful that I gave myself the space to explore, listen to my inner voice, plan and act. It led me to starting my entrepreneurial journey in aspects of my life, which have made me into the true being I am today. These aspects are health, wellness, community service and mentorship.

It was in 2020, during the pandemic, I launched my online social enterprise, which advocates the importance of healthy eating habits, regular physical exercise and provision of a quality supply of all-natural health products to supplement the local citizenry's diet. Through strategic marketing, e-commerce branding and communication via social media, within 6 months of starting up, my customer base tripled. I provided community support by writing healthy lifestyle blogs, conducting mini-training sessions and producing all natural, healthy snacks. This enterprise allowed convenient access to healthy snacks which inspired and influenced people in my community to make better eating and lifestyle choices. I utilized this opportunity to generate revenue during the pandemic, which reinforced my creativity for demonstrating leadership.

In fact, early in my career I managed country-wide volunteers for an NGO (non-government organization), Growing Leaders Foundation, which delivered curriculum to develop and enhance soft skills for at-risk youths throughout Trinidad and Tobago. I specifically left this point to mention in this portion of my story, as it is most relevant here. I was reminded of how much I enjoyed networking and building safe spaces for children and young professionals. I started an online mentoring space to create psychological safety for university students, new graduates, young professionals and females in STEM. Through this platform, I provided coaching, while building confidence and promoting creativity throughout the community.

I sincerely enjoyed this; it made my heart content. I continued to expand my presence online, conducting webinars and speaking at virtual events on numerous topics, which fell under my expertise. Some of these topics included engineering management, carbon capture and storage, green hydrogen and women in leadership roles to name a few. Not forgetting, we were now full-fledged, deep into the COVID-19 pandemic, which changed the world of work and ultimately established a new normal. I embraced all the changes, leveraged my experience and core skills and pivoted. Additionally, I was selected for a mentorship programme with the Volunteer Centre of Trinidad and Tobago in collaboration with the Prince's Trust Fund, which I successfully completed. Wow, how the universe works wonders.

In September 2021, I took on the role as Chairperson of Diversity and Inclusion for the local section of the Society of Petroleum Engineers. I was elated and proud at the same time. It left me thinking, if I did not intuitively listen to that inner voice of mine and believe it was time to move on with my career, would I have gained all this exposure and visibility? I highly doubt it. Introspection and I like to use the term self-reflection, are things that I would recommend to anyone and everyone. Give yourself that fair chance to review the past, project the future and always remember to live deeply in the present moment. To top it off, self- awareness is critical to fully understanding yourself and it just makes it a tad bit easier for things to all fall into place.

## 13.7 YOU DESERVE IT, SMOOTH SAILING

Fast forward to January 2022, my husband and I packed up our lives in two suitcases, literally and travelled across the globe to Abu Dhabi, UAE as I secured a managerial role in an Al ed-tech start-up. This was an ultimate dream come-true. Talk about expressions of gratitude and recognition of the importance of professional diversification and owning your career path. Moving and settling in a new country is fun and exciting but comes with challenges. My exposure to agile project management concepts, learning and development via volunteering and leadership competency made me a favourable candidate for the role. I was thrilled and super pumped to learn, discover and grow.

I'll tell you a small secret, if you look at the bigger picture, my husband and I generated short-, medium- and long-term goals of progressing our lives together into the future. Remarkably, moving and settling together in a forward thinking, innovative country and society was one of our medium-term goals. Indeed, you can go with the flow but from my short experience here on this planet earth, the universe assists with alignment of your goals when you give yourself that self-respect and honour you deserve. Believe that you deserve the best and get out there and rock it like you were born too.

## ABOUT THE AUTHOR

**Liana Roopnarine** is now the PMO lead – Business Transformation at The National Gas Company of Trinidad and Tobago. She has over 13 years of experience in the Oil and Gas industry within major oil companies in Europe and South America. She holds Bachelor of Science in Mechanical Engineering and Master of Science in Engineering Asset Management both from the University of the West Indies in Trinidad and Tobago. She is a registered engineer (R.Eng) by profession and is extremely passionate about Diversity, Equity and Inclusion as she voluntarily chairs the D&I section for the local chapter of the Society of Petroleum Engineers in Trinidad and Tobago.

# *Section III*

## *STEM Reflections*

# 14 Branding and Rebranding throughout Your STEM Career

*Alicia Washington*
Petrochemical M&R Expert, Texas, United States of America

## CONTENTS



When most of us think about branding, we often think of products, goods, or services. Rarely do we ever think about ourselves as a brand. My research on this topic quickly proved that this line of thinking is crippling to our success in the workplace. By definition, branding is the marketing practice of creating a name, symbol, or design that identifies and differentiates a product from other products[1]. Understanding this definition immediately put me outside of my comfort zone. I became defensive and thought "I'm a STEM professional, not a sales and marketing person. Why do I need to learn about branding?" Well, as I started to plan out my career and work towards my next role, the answer to this question became clearer. In fact, I learned not only is branding an essential skill for reaching career goals, but rebranding is also necessary throughout your career if you want to ensure upward mobility in roles that truly fulfill your purpose and passions.

## 14.1 BRANDING: THE DANGERS OF BEING PUT IN A BOX

Upon completion of my engineering degree and landing my first full-time job, I naively thought that my journey to establish myself as a professional was complete. After all, I had the education, degree and the job. What more could I possibly need?

DOI: 10.1201/9781003336495-17

As I continued to progress and develop expertise in those early years, I quickly learned an important fact: I had been branded a "maintenance girl!"

Initially, I struggled to understand what the term "maintenance girl" even meant, but the more I thought about my daily tasks, it started to make sense. I was doing everything from troubleshooting equipment failures, looking for new and better equipment solutions, working on ways to reduce maintenance costs, and physically driving around town to pick up critical parts and get them back to the plant faster than a paid hot shot delivery service. I was definitely all in when it came to making sure the plants I supported ran well with minimum downtime. The fun fact about working in maintenance is that I didn't necessarily choose this career path for myself. Due to my background in mechanical engineering and the privilege of many internships throughout high school and college, maintenance chose me. I found it interesting and exciting enough to continue to invest in learning the nuts and bolts of it all.

In hindsight, I was actually flattered early in my career to be known as a "maintenance girl." That brand said I had expertise and had distinguished myself among my peers as a major contributor in my mostly male-dominated field. It meant that I had persevered and established connections with technically savvy people who saw me as an equal. This was a huge win while in the "gaining knowledge" stage of my career. After a few years as an individual contributor, it became clear that I was not being considered for other opportunities simply because I had become too good at being a "maintenance girl." I do not mention this to be prideful or boastful but rather to help point out the dangers of being put in a box. When I became good at my role, I realized my brand was far too limiting. People in authority over me wanted to keep me in a box for their own personal benefit. I don't believe the intent was malicious in any way, but the limitations placed on my career opportunities at the time could have been detrimental had I not decided to free myself from that box. I needed to establish my own personal brand so I could have the fulfilling career I desired. I had to start making my personal brand a priority.

## 14.2 BRANDING: DOES YOUR BRAND REPRESENT YOUR PURPOSE AND WHAT MAKES YOU HAPPY?

Once branded it can be difficult to completely change how others see you ... though not impossible by any means. Think back to when Apple Computers changed its name. In 2007, CEO Steve Jobs announced that the company Apple Computers was dropping the word computers and would simply be known as Apple Inc[2]. Fast forward to the present day and it is clear that when I think of Apple, I think first about my iPhone and iPad. Apple, Inc. successfully rebranded itself to something that more closely reflected its focus on consumer electronics.

Much like Apple, I too was dealing with a brand that no longer represented my purpose. By this time, I had been leading teams of craftsmen and coaching and mentoring them on soft skills, writing procedures and the importance of our work processes. I found that serving in this capacity was extremely exciting for them and me. I spent hours during lunch breaks and even after hours helping them develop goals and document work instructions that would make routine jobs easier and safer in the future. It was a win-win situation for all and I quickly realized that the next role I aspired to would be people leadership.

I applied for a couple of people leadership roles outside of maintenance and in some cases was not even considered for an interview. It was extremely disheartening for me and for those I worked with who saw my potential. I had people leadership roles outside of the company and I had been doing the role of a people leader by coaching the craftsmen with whom I worked. However, that "maintenance girl" brand was limiting how others outside of my daily work group actually saw me.

In this situation, I consulted a few mentors and decided to make my given brand work for me instead of against me. I began pursuing opportunities for people leadership inside of maintenance. After all, this would make my transition easier by not having to learn a new function and a new role. I eventually landed the perfect leadership role for a site aligned maintenance team. The learning curve was steep having to work with subject matter experts who were extremely seasoned in their careers compared to me. The reward was tremendous in the end. I fostered great relationships and completed many successful projects. I had established myself as a strong maintenance leader thus enhancing my given brand from "*maintenance girl*" to "*maintenance boss.*" In this season, I found fulfillment in my purpose and was truly thriving. So much so that I was asked to move to a different maintenance leader role where expertise was needed. This was the ultimate compliment to my "*maintenance boss*" brand. I was filled with optimism and excitement and felt like my brand had been solidified. I was completely unaware of what was to come.

## 14.3 BRANDING: LIFE CHANGES TRIGGER DIFFERENT CAREER INTERESTS

In 2012, I embarked upon one of the biggest life changes imaginable; I was expecting my first child. It was a year filled with great anticipation and hesitation if I'm being honest. Having spent so many years in manufacturing with the freedom and flexibility to work the long hours, I wasn't sure how this lifestyle change would mesh with the schedule demands I had become accustomed to. I'd already experienced some difficulty in consistently making the early morning meetings. Morning sickness will do that to you! I began to feel the wind shifting in my current career season, but I wasn't sure how to pivot or if I even wanted to. After all, I was so close to finally earning my dream job of managing engineers within maintenance.

My son finally made his appearance in November of 2012 just in time for the holidays. Needless to say, I was on a maternity leave the rest of that year and had some time to adjust to my new life as a mom while thinking about the next phase of my career. I would be remiss not to mention how supportive my team and supervisor were during that time. We stayed in contact as needed, but everyone gave me space to enjoy this season of life and I trusted my team to continue to operate with integrity. It was truly a blissful time personally and professionally.

When I returned to work that January, I found myself sad to be at work and away from my baby, but I pushed through until I developed a new normal. I adjusted my schedule from the coveted 4–10s I worked for many years to a 9–80 schedule.

This allowed me to work shorter days, and since I had to pay daycare by the week anyway, I found those Fridays I went into the office as a good time to catch up on paperwork and connect with my colleagues. Things were going quite well until I accepted a new role and had to step foot into a manufacturing plant for the first time.

I'll never forget the pungent smell of chemicals nor hissing sound of steam lines that I had become completely desensitized to over my years in manufacturing. On this day, I was extra sensitive to it all. So much so that I froze in my tracks. At the time I wasn't sure what was happening, but I now know that my life change had triggered an inner anxiety, which no longer made it fulfilling nor easy for me to continue on my current career path. I was embarrassed and afraid to admit this to anyone for fear of limiting my career and not living up to my then personal brand of "*maintenance boss.*" I mean seriously, a boss is someone who takes and stays in control. She doesn't allow fear or challenges to stop progress. However, at that moment, I knew I needed to start working on rebranding myself if I wanted to continue to thrive in my industry. For clarity, rebranding didn't mean that I had to lose everything I gained in my maintenance career nor that I was quitting. Rather, it meant that now was the time to add to my arsenal of skills and be seen as competent and capable in areas outside of maintenance as well.

In all fairness, I can't blame this epiphany solely on becoming a mom. There are plenty of moms who never feel the need to shift or rebrand. Kudos to those ladies who live their passions each and every day! For me, becoming a mom was a trigger that helped open my eyes to see other possibilities. Prior to that milestone, I was very myopic and focused only on what I wanted and had worked for. Motherhood helped to shift my focus. There was also another factor at play, a HORRIBLE leader who was also a mom who had allowed the male dominated world in which we worked to completely consume her. She was drowning in her relentless pursuit to be seen as an equal and at the top of her game. However, her brand was not strong. In fact, she had jumped around to so many different companies and roles that the only visible brand she had was "difficult to work with." I did not want to be her so I started working on rebranding myself and exiting a toxic situation.

## 14.4 REBRANDING: PACKAGING YOUR SKILLSET FOR A DIFFERENT PURPOSE

As seen with Apple, rebranding is common and effective if done correctly. It can also be scary and frustrating. For me to shed my maintenance brand and open the field of opportunity, I first needed to understand my expertise outside of maintenance. To uncover this mystery, I started reviewing past goals and accomplishments. I also reviewed all my old job descriptions, being extra careful to highlight those tasks I did well and enjoyed. Through this mindfulness exercise, I found a common theme. In every job, whether an individual contributor or leader, I had consistently written and implemented processes and procedures to help work flow smoother and add structure within the organizations I supported. Work processes were a true strength of mine, and as I discovered through many of my own documented results, something that brought value to my teams.

With this newfound information, I began trying to narrow down the types of roles within my company that would complement all the parts of my true brand; from manufacturing and maintenance to work processes. Luckily, the opportunities were plentiful. I set out to do my research on operational excellence, which at the time was simply a fancy buzzword I kept hearing in some of my circles.

As it turned out, operational excellence was exactly what I had been searching for. It was a way to fuse all of my previous experiences and passions into a challenging yet fulfilling role. I was also still managing people in this role except they weren't maintenance technicians. I had a team of truly creative and hardworking employees who mostly made work fun. I had finally broken free of the maintenance brand and was starting to develop my brand as a work process and procedures guru. This rebranding season felt really good!

Just as things were going well, I was forced to pivot yet again and add adult learning to my slate of responsibilities. This was an area I had not had time to explore and with very little warning, I felt like I was being placed in a box again. There was a need that leadership needed to fill and I was at the right place in time to fill it. However, this time I was aware and prepared to protect my brand. I knew the dangers of being placed in a box and although I learned my job and executed well, by no means was anybody going to label me a "*learning girl.*" Not on my watch!

Despite my self-led learning rebellion, I was exposed to many new things in my learning leader role, including grant management and contract negotiations. Although a far cry from the comforts of the maintenance life, I was being challenged in a new way yet still finding solutions. I had also been given the gift of deciding on my brand and how to use it following this role. There were not many people fighting for the role I had and there was a certain peace in that fact. It allowed me time to slow down, load my arsenal and figure out my next move on my own terms.

## 14.5 REBRANDING: PERSONAL GROWTH AWAKENS A NEED FOR NEW PROBLEM-SOLVING

Poet Carroll Bryant once said that "growing old is mandatory but growing up is optional."[3] This is true in our professional lives as well. On the one hand, there are some people who simply want to do a job, get paid and go home at the end of the day. They are not motivated to pursue new challenges nor new opportunities and that is perfectly okay. On the other hand, there are those who thrive on personal growth and professional upward mobility and that is okay also! We can choose to grow old and seasoned in our jobs or we can choose to pursue career growth. I'm definitely the latter and during my season of personal growth, rebranding was at the forefront of my goals.

My last two job roles have been polar opposites with regard to responsibilities, time invested in problem-solving and overall approach to getting work done. Following my stint as a leaning leader, I reverted to my old "*maintenance boss*" brand and took a leadership role in reliability. The hours and physical demands were much better compared to my previous maintenance roles, however, eventually what I learned was that I was simply burned out on the same type of problem-solving. Yes, I learned a lot about budgeting and working with international teams, but at the core of what I was responsible for was still the need to make sure the plants I supported (now all across the globe) ran safely and efficiently. I found myself back in root cause investigations and implementing solutions to solve equipment related issues. It was comforting at first but as time went on, I dusted off my self-reflection handbook once again.

With any job or task, there are tried and true methods of problem-solving and delivering results. The more I stayed in the same box, the more I instinctively committed those tried and true ways to memory. I became like a robot who had been programmed to handle a certain amount of data and then dependent upon the request, effortlessly computed and spit out a solution. Nobody complained about this autopilot approach, and in fact, I received praise for my work. Unfortunately, I am not a robot and as I've grown in my purpose and my life's priorities have shifted, so had my desire to learn and engage in new ways of problem-solving.

In 2007, I received my MBA while working full-time as an inside sales representative. Funny turn of events seeing as how I was offended early on by needing to learn how to brand myself. This was during a time when, without knowing, I was completely rebranding myself and questioning whether or not I made the right choice by earning an engineering degree. There were a few times during those early years when I allowed imposter syndrome to sneak in and convince me that maybe I'd be better at something outside of STEM. I wasn't sure how an MBA would come into play, but it was a personal goal and I was motivated. So, here I was with all this experience and knowledge finally wanting an opportunity to apply my MBA in a commercial role and nobody (and I mean nobody) would give me a chance because I was now branded as a "*reliability leader*." This was frustrating, but I had nobody to blame except for myself. By this time, I knew the importance of building my brand, but I had gotten comfortable and crawled back into my box.

I decided to hire a personal career coach who was excellent at helping me discover and solidify my brand. As it turned out, in everything I'd done from maintenance to operational excellence, I had always been a problem solver. However, my current desire was to find an opportunity to solve different types of problems. My personal career coach helped me to realize that regardless of what my current brand said about me, problem-solving was a universal language across multiple industries and continents. I just needed to go back to the fundamentals and package my skillset into a brand that highlighted my transferrable skills in order to land that commercial role. In May 2021, I was successful!

## 14.6 REBRANDING: PREPARING FOR LIFE AFTER CORPORATE AMERICA

Reflecting on my career journey and how I've had to rebrand myself for the opportunities I wanted, I had the most profound thought following my last role change. Up until that point, all the hard work I put into building and understanding my brand had been solely for the purpose of getting to the next role or job level. Even though I was intentional to seek roles within areas I felt passionate about, I had missed the big picture about branding. I was so determined to stay out of the little box I was once put in that I failed to realize my branding efforts were all done within a larger box, the infrastructure of corporate America. I had not once considered what my brand could or should look like outside the confines of corporate America. This is a big problem many of us fall victim to. Luckily, the problem solver within me sprang into action.

It is hard for many of us to think about our lives after retirement or post the structured corporate environment. After being in industry for 20 years, I still have many unanswered questions about my current role and my future, but after taking two steps forward and then three steps back, I refuse to let this opportunity in the winter of my career get away from me. It's time to get extremely personal with my brand!

Branding experts will tell you that a good personal brand should include seven elements: authenticity, value proposition, story, expertise, consistency, visibility, and connections[4]. This is excellent advice for anyone working on their personal brand but for me, it's all about the authenticity and the story. As I think about my legacy post my formal career, I want the things I am truly passionate about to ooze out of my brand. I've always been a mentor, a cheerleader, a coach, a rule follower, and a right fighter. Whether looking to help the college student on the cusp of switching her major away from STEM or advocating for the new employee who is having a hard time adapting to "adulting" post-college, I am passionate about helping people find and walk in their purpose. I realize in order to do this, I have to operate in spaces where I can make connections and promote my brand to the right audiences.

Within the last year and a half, I have become way more active on social media like LinkedIn and Slack. I've networked virtually with industry professionals and students alike. I also got involved with a magazine curated to show the countless possibilities of a STEM career, called Re.engineer. I've served on Webinar panels, sharing both my expertise and advice, and even written articles to document my learnings throughout my journey. Without giving any thought to why I've engaged in these activities, I realized that I have been working on my brand post corporate America. This is an exciting revelation as I think about the missed opportunities in my past to get ahead of my given brand. It is even more exciting to see how I have evolved as a person and a positive contributor to society and to my beloved STEM profession.

The most important thing to note is that I'm not a "*maintenance girl*," "*maintenance boss,*" "*learning leader*" or "*reliability leader*" as my past brands would suggest. I am in fact **ALL** of those things. The journey of personal branding is more than just committing to a title or slogan. It is about discovering strengths and weaknesses and controlling the narrative of who you are and what you bring to the table. Building your personal brand is about you. When done correctly, your brand will speak highly of you and for you. No matter the career season I find myself in, branding and rebranding myself as a person and a professional is a task I no longer take for granted. I am now intentional about my brand and how I am received in spaces I choose to navigate.

## ABOUT THE AUTHOR

**Alicia Washington** has spent 18 years working in the petrochemical industry. She holds a BS degree in Mechanical Engineering from the University of Oklahoma and an MBA from the University of Phoenix. She currently works as a Mergers & Acquisitions expert on commercial transactions. Previously, Alicia enjoyed a long career in Maintenance and Reliability, including people leadership and asset management spanning four continents. Outside of work, Alicia is passionate about projects that position young adults to lead productive and successful lives. She is an avid volunteer and member of several organizations, but her number one passion is being a mom to her young son.

## REFERENCES

1 https://www.entrepreneur.com/encyclopedia/branding
2 https://www.google.com/amp/s/www.macworld.com/article/183057/applename.html/amp
3 www.carrollbryant.blogspot.com/2012/09/quotes-by-carroll-bryant-vol-one.html?m=1
4 www.entreprenuer.com/article/280268

# 15 From Portugal to Qatar, Exploring STEM Careers in Different Industries

*Vanda Franco*
Petroleum Technology Company W.L.L., Doha, Qatar

**CONTENTS**



"*Life is like a book. Some chapters are sad, some happy and some exciting, but if you never turn the page you will never know what the next chapter holds.*" – **Anonymous**. This is the best way to start sharing my journey in STEM with all of you.

My name is Vanda Franco, I am 41 years old and I'm a chemist (a woman, a wife, a mother, a daughter, a sister, an aunt, a friend, etc.) who had the opportunity to pass through different areas in STEM in Portugal (my country), including marine chemistry, pharmaceutical and petrochemical industries. Today, I'm in Qatar in the lubrication and reliability industry. What will be next? For now, there are too many things to learn and explore in this area!

Everything started at 17 years of age while filling the form to enter into the University – six possible options to fill and only five public universities available near to my hometown with the course which I would like to follow – Nursing. Yes, since I was a kid, I wanted to be a nurse when I grew up!

At that time, it was not possible for me to choose private universities or to select a public university far away from my hometown due to economic conditions. So, as the sixth option, and to ensure no empty spaces, I selected a chemistry course in one of the best public universities in Lisbon – Portugal, as recommended by one of my best friends because her brother was finishing that degree and he was really satisfied.

DOI: 10.1201/9781003336495-18

After 2 months, the list of distribution was published and surprise, surprise, the row with my name was showing chemistry – University in Lisbon! To be honest, I didn't feel sad because I couldn't go to nursing that year due to the final evaluation of high school. Quite the opposite! I was feeling really happy because I had the opportunity to go to a public University near my hometown and my thought was – "Well done, Vanda! You will do the first year of Chemistry and will repeat the national exams of high school to try a nursing course next year!" Now, it is not a surprise to whoever is reading this text – I didn't repeat the national exams the following year and I graduated in chemistry because I loved it!

On this specific course, it was possible to select one of the three areas by the end of second year: education, scientific or technological. I chose technological because theory and practical complement each other and I always need to understand what I'm reading in technical books, published papers, etc. This amounted to many hours of theory but long weeks of practice as well, during the 4 years. The fifth year was an internship to develop skills and finish the graduation.

Different industries and governmental institutes were offering internships. We had to follow a list with classmate names and a final evaluation (average of 4 years) that was built from the highest to the lowest evaluation. I was interested in pharmaceutical or chemical industries; however, my name was in the middle of that list and all offers for both industries were chosen by my colleagues before I had the opportunity to select the internship. I looked at the remaining options and I selected one that I thought "Maybe this will be interesting!"

## 15.1 DEVELOPING THE BASE OF COMMUNICATION

Here starts my journey of almost 4 years as a research analyst in **Marine Chemistry** in one of the governmental institutes at Lisbon, which I never thought about until that day! One of my classmates came to the same institute and we worked together during the 6 months of internship, which was good for both of us to support each other. Today, we continue to be friends and speak time-to-time to each other.

I had to read, and read, and read as much as possible to learn and understand different concepts related with marine chemistry – all new for me! I had the opportunity to improve my English (read, written and listen) and continue developing one methodology onshore to simulate what occurred in sea related with primary productivity – too specific. My first trip offshore occurred here for 2 weeks, and it was amazing to see the sunsets in the ocean. The crew on board was male dominated, and there were five women only.

After my graduation, I was invited to continue in marine chemistry. The governmental institute had a big project to do research in different areas and they required a new member.

I accepted!

Do you know that research is a very-female dominated area?

For 3 years, I worked and learned with amazing co-workers from different departments to analyse and study the contamination of sediments and the quality of water (lagoons, rivers, estuaries and sea), using different methodologies in situ and inside the Laboratory. Yes, I also had some colleagues who in the beginning were not the best person to me, not passing all of the information and creating friction because they thought that I was competing with them. We spoke and I explained that I was learning with them, I was not there to replace them. There is space for all if you want! Today, I continue sending birthday wishes and Christmas cards by email to some of them.

### 15.1.1 Communication and Respect Is the Base of Each Relationship, Personal and Professional!

To grow and achieve different levels in this area – research – higher education was required, so I decided to continue investing in my education by doing a master's degree in biology and management of marine resources. It was not easy, working and studying at same time is hard and exhausting! No social life at all during the first year! But with the support of my family, friends, colleagues and teachers, I did it and with success. I finished my master's degree in 2 years! During this period, I had the opportunity to attend different conferences, participate with poster presentations and contribute to different scientific papers. I was so happy about these achievements!

Relieved of the pressure, I could finally start thinking about my next personal goal – purchase an apartment! However, 1 week later, I received the information that the budget for the project I was working for was reduced and they will not be able to keep me on staff. At that time, they didn't even know when another project or injection of capital would come to the project. Another complicated decision –

> "What to do now? I spent 2 years of my life and invested one part of my savings to have my Masters degree and be able to grow in this career! To have my own apartment and be independent! Now the future is uncertain!"

I had an honest conversation with my superior – an extraordinary researcher – to check the possibilities and I decided to use my knowledge, short experience and studies to move on and started searching for a new job. My superior tried to keep me there until a new project became available, but it would be unpaid work – how can we manage our life without a salary? I left the door open to go back one day, if necessary. The project to purchase an apartment was postponed.

## 15.2 NAVIGATING NEW PATHS

Here starts my journey of almost 4 years as a quality control analyst (QCA) in one recognized **Pharmaceutical** industry near Lisbon!

Three months (not too long) was the time that I spent sending CVs and having interviews until being selected and hired by the pharmaceutical Industry. On my third and last interview, I met a lady who was also waiting for her final interview. She was a teacher in a high school for a few years and she told me "I'm so nervous! I would love to be selected to work in the pharmaceutical industry." You know what? Both of us were selected and we started on the same day. Another goal achieved! She is working there today and we are really good friends.

*Do you know that, once again, the Quality Control and Assurance departments are a very-female dominated areas in most industries?* I can say that in both teams (one pharmaceutical industry, two different locations) there were 100% women including the management team.

Considering that I didn't have an experience in a pharmaceutical laboratory, I started as a QCA for package and production departments, doing weekly shifts – day and night. Microbiology control was also included in both departments to prevent and control possible contamination from air. It was the best that could have happened to me because I learned everything I could from all departments, from raw materials to the final packaged product ready to be delivered to hospitals, primary health centres, pharmacies, etc. It was not easy, I can say, but it was worth every hour. Personally, I had the opportunity to work with people from different economic classes and education levels, which encouraged me to do more and better for them.

In 6 months, I joined the laboratory team with an open mind and had a different idea about what was happening outside the laboratory. My co-worker, who became a friend, joined the laboratory on the same day – different areas. We were very happy and supported each other. Before we joined the laboratory team, we trained two ladies who were hired recently to replace us. We were given 1 month to prepare them by explaining all steps and tasks in each department – production and packaging. Another friendship started with one of the new colleagues, which lasts until today.

The laboratory facility was big, very well furnished with the latest equipment technology, totally different from the laboratories in the university and in the governmental institute. A young team with so much potential was waiting for us, however, under high competition and pressure. They did not have time to complete their tasks, train us and transfer the required information, so we had to apply our proactivity and start doing by ourselves – read each procedure and follow it. Whenever a doubt appeared, we clarified with the respective experienced colleague who was supporting us for raw material analysis and final product stability. The team was divided by areas: raw material, final product, product stability, microbiology and research/development.

"*Do what you can with all you have, wherever you are.*" **Theodore Roosevelt** has stuck with me! One month after joining the laboratory, I received a call to meet the quality assurance manager. It was Friday evening, finishing my shift, and on the way from the laboratory to the head office in a different building I was thinking "What is the reason for a meeting on a Friday evening? Why did my Supervisor not mention the subject to me?" I got there, sat down and waited for the QA manager to start speaking. She asked me "Do you know why you are here?" "Should I?" I replied. After my negative response, she explained to me that one QCA was required to be moved to the other laboratory to support the respective team to train them on the latest equipment technology which they were receiving. The management team did the evaluation and I was the one selected to be moved due to my skills of fast learning and adaptation to different teams – starting the following Monday. In the beginning, it was strange for me because I thought, "How can they evaluate me in such a short period and know that I'm the right person to go to the other laboratory? There were co-workers with more experience than me and working there for a longer time."

After discussing some details and learning where the other laboratory was (area and address only), I accepted, once again, to change and start from zero: new Laboratory, new team and new managers. I went back to the laboratory to say goodbye and pack my stuff. During the conversation with my colleagues, I realized that some of them had the same meeting but didn't accept to be moved due to personal or professional questions. This was the reason why, a recent employee, received the request to have a meeting and be moved. I was not the first option!

*"Today I will do what others won't, so tomorrow I can do what others can't!"* – **Jerry Rice**. This was my mantra for this new position. During the weekend, I did my preparation – drove 25 km from my hometown to the city where the other laboratory was located, to learn the way and ensure I wouldn't be late on the first day. From outside the building, it looked totally different – older than the previous one. My thoughts were, "What about inside? The production and package area? Be calm, you will see everything on Monday."

Monday arrived and I went to the other laboratory very early in the morning. The laboratory manager was waiting for me. I was very well received. We started the visit in the laboratory – smaller, older, less material and few equipment; however, it had everything that was needed to perform the analysis. The team was older and with one-third of the number of QCAs. The production and package departments were also smaller with employees who were working there for 20 years. I thought "I cannot imagine myself working in the same place, with the same co-workers, for 20 years!" But, in this industry, they were like a family, with some not so good days as a normal family. Here, the research/development area didn't exist and the microbiology area was very small.

One of the best things was only one working shift and it was during the day. Work on Saturdays only if really necessary. One of the worst things – too much traffic, and if we tried to use public transportation, it took 2 hours 30 minutes each way, which means that you would spend 5 hours of your day inside the bus, metro and boat. I decided to go by car.

The management of a small team should be easier, don't you think? No, it is quite the opposite because each person needs to know how to do everything, from the raw material to the final product passing through microbiology and so on.

Remember, I was shifted to this laboratory to provide technical support with the latest equipment technology? In the end, I was the one who learned so many things with all of them: how to perform some parameters faster, troubleshooting of equipment, how to work with old technology, different software and much more. Great experience with experts in the field and without feeling the competition of a young team, they shared their knowledge with me every day.

Two years later, I felt stagnant with the routine activity. I would like to do more, do something different inside the pharmaceutical industry to contribute to company development. I had a conversation with my new superior to find out the possible opportunities and how I could grow and develop in this career. She passed this to the top management, and they had a meeting with me to explain that it will be difficult for a chemist to grow in QC & QA in the pharmaceutical industry. To grow, I would need the degree of pharmaceutical sciences, which meant going to university again and quitting my current job because the schedules were not compatible. Besides, there were no open vacancies in the other departments and the management team couldn't ensure a better position/different function for me until I finished a new degree, so I didn't proceed.

Quitting my job was not an option, I had bills to pay including rent. Yes, at that moment, I was already living in my own apartment together with my boyfriend (current husband). Going back to marine chemistry was not possible because the projects continued without financial support – I knew this because I kept contact with some of my ex-colleagues.

I decided to start looking for a new job where I could apply my knowledge, learn more and help the company to achieve their goals while growing as a chemist. It took me 1 year to get that opportunity, but it was possible. On the same day, I informed my superior in the pharmaceutical industry that after passing through a process of five interviews, I had been selected by a petrochemical industry and I should start in 1 week. It was not easy for them and they tried to keep me there, but they didn't have anything to offer to me, the same speech as 1 year before. After discussing it with my family and my husband, I received their support. I accepted the offer and moved to another city during the weekdays – rental apartment. During the weekend, I came back to my hometown.

Nowadays, it is different in Portugal, there are chemists in different departments of the pharmaceutical industry and laboratory supervisors who have a chemistry degree only.

## 15.3 STARTING FROM EXPERIENCE NOT FROM SCRATCH

Here starts my journey of 6 years in one **Petrochemical** industry in Portugal, 150 km from my hometown, which was under pre-commissioning and where I had the opportunity to develop and grow from an analyst to a laboratory supervisor in 18 months.

I would like to share a curiosity about the process of five interviews I underwent to be selected, where in each of them there were three repeated questions made by different people – human resources manager, laboratory manager, production manager and chief executive officer:

- What do you think about starting from zero as an analyst and working by shifts, when you are already 30 years old and have laboratory experience?
- Are you comfortable to work with a team where you are the oldest one? The youngest is 18 years old.
- The education level of the team is technical. Are you comfortable working with them, at the same level, receiving the same salary, considering you have a master's degree?

For me, it was easy to answer all of the questions because I had already experienced the "start from zero" twice, I worked by shifts and I got used to working with colleagues from different ages and education levels. We can learn so many things from each one of them, personally and professionally. We can share our knowledge, develop together and achieve more.

The first two days were filled by transversal training, knowing each department and manager. Learning about the plant and the process/production. We were a group of six in total having this introduction, four women and two men, recent employees selected to be a part of four different areas – laboratory, IT, supply chain and logistics. What a group! We felt like we would know each other for years. Until today, four of them continue working in this company and two of them became really good friends of mine with whom I continue sharing my achievements, my happiness and my not so good moments.

At that time, the percentage of women in operations, maintenance and SHE departments was reduced: four women were part of the operations department – process engineer, supervisors and operator; one woman in maintenance and one woman in SHE departments. On the other side, the HR, logistics, supply chain, quality were dominated by women. Really interesting!

In one of the breaks of the first day, the laboratory manager took me to visit the laboratory and introduced me to the team that was waiting for me. The facilities were huge and really good, filled with the latest equipment technology; however, all of them were covered and protected with plastic – it was strange especially because I knew that the plant was already under pre-commissioning. I was well received by the team, which was composed of 11 analysts (five men and six women), one senior analyst and 1 manager, both women. I would be the 12th analyst. As per the plan, there were supposed to be five teams of two analysts rotating by shifts (day and night) and one team of two analysts in day shift only – rotating per year.

The following week, I started my 2 weeks of training inside the laboratory with the laboratory manager to learn each procedure, see all equipment (20) and the different software (10). There was a specific room with high technology and the most critical for me at that moment because it was the first time I was going to work with these technologies – inductively coupled plasma (ICP), atomic absorption spectrometry (AAS) and energy dispersive X-ray (ED-XRF) – I had the theory; however, I never had the opportunity to work with it.

Regarding the team of analysts, they were divided in two sub-teams determined by the date that they had been hired, the respective trainer – senior analyst or laboratory manager – and based on the group of equipment on which they had received external training. Once again, this was strange for me because from my previous experience, I learnt how to work with all equipment and be autonomous.

During the training period, I had some conversations with the team to find out a little bit about each member. I realized that they knew my entire CV! Yes, the senior analyst had informed them about my education level and previous experience. They spoke about themselves and about their relationship with the senior analyst and manager. What I could expect in the future regarding the behaviour of each one.

Third week started and the plan was to receive training on each equipment from the "Critical Room" as I called it. The trainers were three different analysts as per their knowledge and different development. The senior analyst didn't give me any training until that moment – strange, no? In the middle of the week, we heard some noise inside the manager's office … she was shouting with the senior analyst. It was a complicated situation and I could understand the previous conversation that I had with analysts – this was a recurring situation, the laboratory manager used to shout with all of them and human resources was not acting.

Fourth week – I received a call from human resources to have a meeting to propose me to be senior analyst! Yes, in less than 1 month, I was starting a higher position in my career inside the petrochemical industry, which would be confirmed after 6 months of evaluation. I realized that I was selected for this position from the beginning, not for analyst. No one explained to me what happened with the other senior analyst, I was just informed that she was going to a different department. They requested me to be patient with the laboratory manager and try to calm down the situation inside the laboratory – they were trusting me and trying to improve it.

*"If somebody offers you an amazing opportunity but you are not sure you can do it, SAY YES – then learn how to do it later."* – **Richard Branson**. This echoed in my head during this phase! I can say that two or three analysts didn't receive this information with a smile, quite the opposite! They cried and complained because they were really connected to the other senior analyst. They were good friends and, unfortunately, they were mixing personal and professional relationships. They were feeling afraid and thinking that "everything will be worse from now on". It was not easy to convince them that I was now their senior analyst. A few months later, the situation changed a little bit because they knew me a little bit more. Until today, I continue to keep in contact with two of them and I'm really happy about that.

Another 2 weeks passed and I understood what human resources had requested of me. It was really difficult to work with a superior who did not support you, who shouted at you because of everything, pointed the finger at you every time you failed, not giving you the opportunity to expose your ideas and suggestions and so on. The threats were constant in relation to dismissal. Unfortunately, the woman who was supposed to be my mentor showed me everything that a leader should not be! She was practicing bullying with each one of us.

It was time to speak with the human resources to expose the situation and request their support. Some analysts did the same previously but nothing changed. They listened to me carefully and the solution presented by them was to provide training to us – *Personal Effectiveness*. Later, I noticed that our laboratory manager didn't accept the training. In the end, I can say that it was better to receive the training without her because we were relaxed and we had the opportunity to understand ourselves better and find out a bit more about each colleague, as an individual and as a team.

*"While you cannot control someone's negative behaviour, you can control how long you participate in it."* – **Anonymous**. Five months left to press the "start" button of the plant and all equipment inside the laboratory was still turned off! We were not allowed to do anything, only read the procedures and translate from English to Portuguese. All analysts had intensive training provided by suppliers; however, they were not practicing. How was this possible?

Our laboratory manager had gone on a vacation for 2 weeks. I spoke with the operations manager (from the United Kingdom) to explain the situation, and after his approval, I removed all plastics from equipment at first opportunity, turned them on and sent one team to each equipment to create the methods, prepare the required standard solutions and start the calibrations. There were many methods to be created, many standard solutions to be prepared, many calibration lines to be tested and retested. At the beginning, they were afraid of it because they had forgotten some steps since they didn't practice after receiving the intensive training. I told them to relax and start slowly, one by one. Read the manual and work instructions. I worked 14–16 hours per day, but everything was done smoothly and in a proper way.

Our laboratory manager came back from vacation. There was a big discussion as you can imagine! Her behaviour didn't improve after vacation. She was supposed to rest and relax, but she came back worse than ever. And day by day we were feeling sad, without proper psychological conditions to continue. I tried to speak with her to explain the situation and she changed ... for a short period only! She is a clever woman, however not showing the skills to manage a team and work in a group.

I can say that I didn't quit because of all of them, my team!

One month prior to start the production, we initiated the shifts, including me, because one analyst was missing and recruitment was taking place. All patented methods were followed as a recipe to prepare a cake in the kitchen. Most of the equipment was ready to receive the samples from raw material, from production to follow the process and the final product; however, two critical methods were not correct and not giving accurate results for the final product. As I mentioned previously, we were not allowed to give our suggestions so no one was thinking about this issue ... what could be the root cause?

Two weeks left for the big day and we received a visit from one of the experts in a petrochemical laboratory with a wealth of experience in management, a man from the United Kingdom with more than 30 years of experience and lots of knowledge. I learned more from him in 2 weeks than with our laboratory manager in 6 months. After having a meeting to discuss all pending actions, I was allowed to share my thoughts about the methods with inaccurate results to find the root cause. I did some tests and we realized that the methods were not correct regarding the standard solution preparation and the concentration of some solutions. I worked 18 hours per night/day during 1 week to assure that all methods would be ready before we received the first sample from the plant. The new analyst was hired and she joined us. She received the training from the laboratory manager while I was supporting the team.

The big day arrived and we were all nervous and excited at the same time – 150 employees in total, from different nationalities (Portuguese, Spanish and English). Samples arrived every hour to the laboratory, without being able to analyse all the parameters as per the request. Fortunately, the expert visitor stayed with us for another week. He gave some tips to coordinate everything and so we do not panic.

This month, I received confirmation from human resources that my evaluation was good and I would be the senior analyst. What a month! It didn't start well but ended up excellent! Another goal achieved.

One month later, the English operations manager left. New managers were selected inside the company and they were introduced to us – some I already knew from the transversal training. They used to work with our laboratory manager previously, so they knew everything that was happening. They knew that her behaviour would not change, unfortunately.

Considering that the laboratory was working 24 hours to support the production, I was on call during the night. Not easy, at all! The laboratory manager was not picking up the call as supposed to. There were too many problems and human errors due to lack of experience and tiredness of the team. I spent a long time inside the laboratory. For almost 1 year, I didn't know the faces of the supervisors from the plant, I knew them by phone only.

My vacation period arrived, finally, after 9 months working "without breathing"! The feeling was ambiguous because I really needed to rest, but I knew that by not being in the laboratory the team would have to have more contact with the laboratory manager because she would replace me during my annual leave.

As soon as I returned back from vacation, I received the information that the annual evaluation of the team had occurred without my knowledge and my contribution. They were not satisfied with this situation; however, I was not able to change it, it was a top management decision. I was evaluated as well, as an individual and as a team. Some improvements to be done, as normal.

Never stop improving yourself!

During this evaluation, there was a question related with which team shall be moved from day/night shifts to come to day shift only during the next 12 months. It was suggested to select the one with more experience to support me and allow me to be available for different tasks. I appreciated management's concern; however, I requested the team where one of the analysts (a man) was showing some flaws and needed to improve. Yes, I knew that I would have a lot of work ahead of me, but it made sense to me. It was accepted. It was not easy, to be honest, because I was pushing on him, and at that moment, he didn't understand that I was helping him to grow. Six months later, I was so proud of him! He was able to work with all equipment and with each software. Nowadays, we still have conversations about this period and he continues thanking me for what I did.

"*A leader is someone who can help the other one to grow and develop new and unknown competences.*" – **Jose Pinheiro**. Only 1 year after I joined the company, I started having a social life in this city, enjoying the best that this place could offer to me during the weekdays. I was living near the beach – 10 minutes walking distance! In the winter, it was too cold, but during the rest of the year, it was possible to do exercise outside, breathe freely and watch amazing sunrises and sunsets. I was able to have a good time with my colleagues who joined the company on the same day and tried to relax after work and too many hours inside the laboratory.

Another 6 months passed, and during this period, the behaviour of our laboratory manager didn't change. We were really exhausted by such actions and reactions. She was having the same behaviour as all co-workers from different departments, especially if their function level was below hers. The operations manager noticed that and I received another proposal – to be the laboratory supervisor, a new position to be created, and manage the team! Yes, in 18 months, I had the opportunity to grow and develop from an analyst to a laboratory supervisor, as a chemist! I was managing the laboratory and a team of 14 analysts in total.

## 15.4 BECOMING A LEADER

"*I am not lucky. You know what I am? I am smart, I am talented, I take advantage of the opportunities that come my way and I work really, really hard. Don't call me lucky. Call me a badass.*" – **Shonda Rhimes**. This proposal was made to reduce the contact between the current laboratory manager and the team. In fact, the laboratory manager position was eliminated and she received a proposal to do laboratory technical development and research to support the engineering department. At this moment, I was reporting to the operations manager directly. Everything started getting better, the team was tired but happy. It was possible to see big smiles on their faces while working. It was not easy to bring them up again, but we made it together!

A few months later, during a conversation with some of my co-workers while having dinner, the subject of sharing an apartment came to the table. Each one of us was living alone, in different places. So, we thought that it could be a good idea to save money and have someone at home to speak with when we arrive from work during the week. We did it, there were four of us so we looked for a villa with four bedrooms and a big space. It was a good time, I can say! We lived together for more than 1 year and we learned a lot about each other personally! Later, one of them got married and we looked for different places again.

Once again, the operations manager changed, the current operations manager was leaving! This time, an Italian gentleman with a lot of experience within the petrochemical industry joined us. He was introduced to me, and on the same day, I started reporting to him.

During the next 2 years, this industry had ups and downs due to market constraints related to crude and oil & gas fluctuation. I used the standby periods to train our team in all the equipment, to teach each one of them to be autonomous and eliminate the lack of knowledge. It was also possible to reduce the calls during the nights and weekends because they were able to do the basic troubleshooting when necessary. Internal and intensive training was planned with final evaluation. I shared with them all of my knowledge and experience. As a trainer, I also received evaluation and was able to improve.

*"Train people well enough so they can leave, treat them well enough so they don't want to."* – **Richard Branson**. During the shutdowns, the laboratory team was able to support the maintenance team by being confined space watchers, after receiving specific training from the HSE department. Here, we had the opportunity to contact colleagues from different departments, understand their tasks and learn more about each machine. I also had the opportunity to be a SUSA (Safe & UnSafe Acts) auditor to check ongoing activities and try to improve the UnSafe Acts. The Safe Acts were also registered to congratulate the co-workers for the good practices and to be used as an example.

Another evaluation was performed. This time I could evaluate all analysts by myself, without any intervention. I was also evaluated, and during the meeting, the new operations manager told me "Vanda, you know I'm a macho man, and this is the first time in my life that I'm giving the highest evaluation to a woman. I'm expecting that you can continue with this good performance." I was really happy because all of my efforts were being recognized. In the end, before I left his office, he added one request "Please, do not get pregnant during the next year because I need you in the Laboratory, available at any moment!" Wow, I didn't know what to think or to say … my superior was invading my personal life. This shouldn't be requested of any woman at any moment, we have the right to decide when we want to have a child.

Suddenly, a 6-month layoff came and the laboratory team was sent home. It was a big concern for all of us because we didn't know what could happen after 6 months, if we could return and continue working or if some of us would be fired. One of my concerns was – "Will I have all of the assistants available when we return? Do they search for another job?" Fortunately, none of us were fired and all came back from layoff; however, colleagues from different departments were invited to leave the company due to reorganization.

The laboratory had to be prepared similarly with commissioning. Everything went smoothly. Unfortunately, two analysts left the company to go to a different city, a few months after our return. Time to improve myself, once again, to continue developing the laboratory and supporting our team! I decided to do the postgraduate diploma in laboratory management to increase my knowledge in different areas such as human resources, financial, quality management systems, method validation, etc.

*"Before you are a leader, success is all about growing yourself. When you become a leader, success is all about growing others."* – **Jack Welch**. My salary was not revised after receiving the postgraduate diploma, and compared with the operation supervisor, it was very low for the same function, tasks and responsibilities. However, there is a large gap in Portugal between salaries of men and women in the private sector. Unfortunately, this still continues nowadays.

One year later, after finishing my postgraduate diploma, my husband received a proposal to be transferred to a different country – Qatar – continuing to work within the group related with chemical cleaning and be the operations manager of a joint venture. After analysing all the aspects, and thinking that 3 years ago he rejected a similar proposal to go to America due to different reasons, we decided and he accepted the proposal. At that time, we discussed that he should go and try. Two months was the time between taking the decision and travelling to Qatar. I told him, if he adapts I would think about going to Qatar too, but with one condition – having a job and if possible as a chemist. I know myself very well and leaving everything in Portugal to make a home in Qatar was not a choice, at all!

Since the company group was international, my husband used to travel around the world and stay for long periods outside Portugal – maximum 2 months for each project. This time it would be different! He was going to live in a different country, be a resident and be alone for some time!

The production within the petrochemical industry continued with some fluctuations, and in 3 months, another layoff came. The laboratory team was sent home, once again, except me. It was requested to keep me working to support the technology department – yes, the laboratory was moved from operations to the technology department along with process engineering. I also had a new manager to whom I should report. During this time, I had the opportunity to work with different colleagues and learn so many things about the plant – equipment, functions, each step of the production, sampling points, correlation of analysed parameters, etc.

In the middle of the layoff, 5 months after my husband left Portugal, I took the flight to Qatar and stayed there for 2 weeks "on vacation". It was possible to see that it was a totally different country from what we got used to reading in the news about Muslim countries – a very small peninsula in Persian Gulf with around 2.8 million people where only 15% are locals. What a mix of nationalities, cultures, religions, habits, food, costumes, etc. There were amazing landscapes and a very developed city, a lot of construction, not only desert! It was also, too hot and humid; however, there was air conditioner everywhere. During these 2 weeks, I met some women and they explained to me that there was a small percentage of expatriate women working in Qatar, especially in the industries/plants. Most of the expatriated women were working as admin, teachers and nurses. In fact, most of the nurseries were looking for teachers to start on the next school year. They explained that it would be easier when

I was already in Qatar to be considered as a resident and I would not need a sponsor. Each expatriated – man our woman – needs a sponsor to be in Qatar, and in all the other GCC countries, from a company or from family. In my case, my husband would be my sponsor.

I did some research to find out the possible websites and where to search for opportunities in Qatar within industry – there were too many requesting registration and to fill their own forms with all details from the CV. It took almost 2 hours to fill each one and the last field was to attach the CV! This did not make sense! But it was the only way to be registered on their systems and receive the current and new vacancies in Qatar!

Time to go back to Portugal, after "vacation". It was not easy to leave my husband again … 6000 km separating both countries! As soon as I arrived, I received an email from one company in Qatar requesting me to answer a few questions and to have an interview for the position of a chemical analyst in a concrete laboratory. Due to some problems from their side, the company cancelled the offer and the interview didn't occur.

Back to petrochemical to continue working during the current layoff.

There were no answers to my emails relating to any vacancies in Qatar. To be honest, I was submitting my CVs only to chemical and laboratory-related vacancies. The first interview was for a teacher in one of the nurseries, through one of the Portuguese women I met during my vacation in Qatar, I was selected and approved. Yes, in Qatar, it's easier for a woman to start if we have contacts. Next step – negotiation of conditions and coordinating the starting date. I took the decision based on the thought that "once I'm in Qatar I could more easily search the vacancies and inform them that I was physically there and available for an interview." The important thing at that moment was to go to Qatar as soon as possible.

Time to inform the manager that I would resign and leave the company. I explained everything! It was not easy but he accepted. The general manager requested a meeting with me, as well. After our conversation, they requested me to select one person from the team of analysts to replace me. I gave two options: male and female with respective justifications. The final decision came from the top management – the male was selected! At the first opportunity, they chose a man to lead the team! It was now time to inform the respective person and our team. Considering that we were still in layoff, a special meeting was requested and all of them came to the laboratory. They were not expecting the subject of my leaving. I explained everything and they understood! We also announced that one analyst was already selected to replace me. On the following day, I started the training, passing the "folder" to my colleague. I gave 2 months to the company and each day was used to train the analyst who would replace me. I did my best and no point/subject was unsaid – intensive training. During these 2 months, the technology manager told me several times – "You still have time to not resign and continue with us!" It was good to know that the door continued to be open, but the decision was taken – time to leave and start a new life! The last day was not easy … OK, the last week was not easy, I can say! There were too many memories, good and not so good!

One week to go to Qatar and say goodbye to my family and friends until next vacation! Another difficult week, full of anxiety, expectations and already missing all of them! The day to travel arrived and I had requested my family and friends to not come to the airport, I didn't want to be nervous! Of course, some of them didn't follow my request and appeared there as a surprise. It was really good to be with them for another hour … but I left crying like a baby!

## 15.5 MOVING TO QATAR

Here starts my journey in Qatar till today, 6000 km away from my hometown. At 36 years of age, I started as a teacher in one nursery, and in less than 1 year, I was joining the oil & gas services provider company – **Lubrication and Reliability industry**!

Arriving in Qatar! It was too hot and too humid, as expected! There were so many places to see and things to learn! I was really excited!

One week to prepare everything, including documentation, before starting the new job as a teacher for the first time! The main language for communication in Qatar is English! In Portugal, we speak Portuguese from North to South, English is spoken only if it's really necessary. I was glad that I had developed the writing, the reading and the listening during university and in all my jobs. To be honest, my speech was too technical due to research, I had to improve and develop it to communicate on a daily basis. Nowadays, I speak Portuguese at home only, at the end of the day, or when texting to my Portuguese family and friends.

First week – everything went smoothly, introducing me and my assistants to the students (3–4 years) and parents, understanding the routines of each one, preparing the lessons and so on. All co-workers were women, no men in the nursery.

It was a mix of cultures, different education, different needs, different requirements and expectations from each parent. I loved to teach the kids and play with them; however, the behaviour of some parents was affecting me because they were not respecting us; myself and the assistants, and unfortunately, I was not allowed to say anything – management decisions. I was not sleeping well, I was not feeling happy, and unfortunately, I didn't adapt. After 4 months, I told my superior that I was resigning. I was so disappointed because I thought that I could do and adapt to anything. I always pushed myself to the limit. It was not mandatory because I did not have a contract; however, I gave 1 month's notice to the nursery to find and select a new teacher. Two weeks was the time devoted to passing all the information and details related to each student. During this time, I was feeling relieved and sleeping better. The positive thing – I was able to help one of the Arabic students who was taking a longer time to develop – in a few weeks, he was saying some words in English and interacting with all the classmates. They just need time and patience. This quote really spoke to me during these times; "*If you try and fail, congratulations! Most people won't even try.*" – **Anonymous**

## 15.6 NURSING MACHINES INSTEAD OF HUMANS

The first thing I did after leaving the nursery was to update my CV and prepare a short version of one page only with relevant information. I requested the support of one of my friends who was living outside Portugal and had a lot of experience in preparing CVs and analysing the same to hire personnel to her team.

The next thing I did was update my LinkedIn profile and insert a summary with keywords to be found easily when the companies are searching for specific profiles.

The third thing was to start my search for a new opportunity and this time I was honest with myself – it needs to be related to my area. I will not compromise this! I submitted CVs to different offers, on different websites! I printed some CVs and delivered them to the reception of different companies. And surprise, surprise, I received calls and had interviews with some companies where I had delivered the CV by hand. I didn't receive any response from the CVs submitted by email related to open vacancies.

After 5 months of submitting CVs, and having some interviews, I reached the final step of the process to be a technical sales of equipment for a laboratory. While discussing the offer and the contract I was informed that I should start after Ramadan only, in 1 month, so the team would be available to give me training and for all laboratories to return to their normal function. During the month of Ramadan, the working time is reduced, especially for governmental entities. I left their office with information to wait for their call to confirm the starting date and to sign the contract for 1 year.

I called my husband, and after talking about the conditions to be technical sales, he told me that the oil care department of his mother company in Qatar was looking for a chemist to be the focal point between the third-party lubricant analyses laboratory and their oil & gas/power plant customers – urgently! He asked me if I was interested to have an interview with the operations manager and sales general manager to know a little bit more. Why not? I never worked in oil & gas, but it could be interesting.

Two days later, I had the interview. Everything was explained to me. For the first time, I had a practical interview with questions to solve some real problems. I left their office with a clear perception of what they were looking for and their expectations. Even without background related with mechanical, oil condition monitoring, lubrication and reliability, etc. In 2 days, I received a formal offer to be discussed – if my answer was yes, the contract would be written. They were giving me an opportunity considering my experience in the laboratory. This time was not to work inside the laboratory, but to work closely with different laboratories to have the lubricant analysis done in a proper way and on time.

After analysing both proposals, discussing with my husband and checking the opportunities in the future with each function – what I could learn more and how I could support the company with my knowledge and experience – I decided to accept the offer to be technical support for oil care services provider, even with a lower salary when compared with technical sales proposal from the other company. In 1 week, I was starting my new function, with 4 months' probation period, and until today, more than 4 years later, I know that was the right decision!

On the first day, I realized that our team was male-dominated, only one woman was administrative and now me – 2 women and 21 men, 6 nationalities! Wow! For the entire company, the percentage of women was 13% only, 75 women in a total of 577 employees – 35 nationalities! WOW! Each nationality has a specific accent, which is perfect to train the listener.

On the same day, two of my colleagues left on annual leave and I had to replace them for one month, doing everything related with lubricant analyses – logistics, sampling coordination, analysing the sample reports, quotations (commercial and technical offers), submitting proposals under tender invitation, etc. Gladly, I had the support from my manager, our admin and the whole team. It was intensive training and I could feel a big welcome from their side!!! It was their first time working with a woman in this area of lubrication.

On the third day, I was wearing coveralls and going to the workshop to learn each equipment and about all the services that the company was providing. Within 2 months, I was going to some of our major customers to have meetings related to technical clarifications, together with the business development manager and one of our partners. Some of them were surprised to see that I was a woman since we were having conversations by email only.

This was the first time I heard about the concept of Varnish, how to measure and how to mitigate it. Another intensive training!! One of the comments of our partner was *"If you want, Vanda, one day you can be the Queen of Varnish in Qatar!"* – I laughed because it was a funny comment. I was just starting! Four years later, I'm not the Queen or even the Princess, but I'm feeling confident, and clients trust me to support them and solve the issues related with lubricant contamination that each machine presents. In fact, we are the doctors of their critical machines, we take care of each one as a patient: we analyse the blood (lubricant), we do a check-up on each part of their body (symptoms of the machine), we consult the current and historical diseases (issues and corrective maintenance) to find the possible root cause and recommend the best medicine (solution) to be applied and solve the problem. Sometimes, it's necessary for us to go to the operating room (on-site) to apply the respective solution – preventive or corrective.

Within 1 year, I was able to go to the field with my colleagues! I received all the training before going inside the plant. I was the only woman in the classroom for all trainings including the certification in competence in condition monitoring – CAT.I – BINDT certification in the UK sponsored by the company. For the first time in the plant of one of major oil & gas operators in Qatar, I was missing this! Only on site, we can learn more, we can understand better what the issues are. Even with the large number of employees and sub-contractors, I was the only woman – a very-male dominated industry, especially in the GCC countries. During lunch time at the cantina, I saw a few ladies from the admin department, but they were in different areas. I had my lunch together with my team, in the common area, where it was possible to have a non-formal conversation with each contact from our customer. In the end, they were the ones who spent their lunch time knowing me – curiosities about a European female working in this industry. Nowadays, most of the focal points from the different areas know me or speak with me by phone or email. Even today, only one man refuses to discuss with me by phone because I'm a woman. I'm feeling comfortable in the field and I have strong support from our team and company. I'm autonomous in most of all my responsibilities.

Considering that the weekend in Qatar is Friday & Saturday and that most of our suppliers and partners are from America and Europe, I use to work a little bit on Fridays to follow up by email the pending orders and requests, avoiding the three consecutive days (Friday, Saturday and Sunday) without communicating with them and without response to our customers.

## 15.7 PREGNANT IN A FOREIGN COUNTRY AND DURING COVID-19

"*Success is no accident. It is hard work, perseverance, learning, studying, sacrifice and most of all, love of what you are doing or learning to do.*" – **Edson Arantes do Nascimento** known as **Pelé**. These words ring true for me today! After finishing my first oil sampling campaign in the field, I knew I was pregnant – few weeks only! As per the doctor, everything was OK and I could continue my normal activity. I went to the field until I was 5 months pregnant and I wore the coveralls until I was 39 weeks pregnant – it was a healthy and energetic pregnancy. The most difficult thing was not being able to go to Portugal when it was planned. I spent my entire pregnancy without seeing my family and friends, only by video-calls and pictures. The best was to have four friends in Portugal pregnant at the same time and be able to share all doubts and concerns – everything was normal! New plans – go to Portugal during the maternity leave to introduce our son to the family and friends.

One month before I started my maternity leave, we were allowed to hire another person to replace me in some of my tasks – it was an internal process and we selected a woman! A chemical engineer! She was working in the company for 3 years as an admin in a different department! She was really happy to be transferred to our department and be able to work in the field! She joined in 2 weeks, and now, we are two women in a very-male industry in Qatar and she is giving excellent assistance to me and the whole team!

Our son was born on the day I completed the 40th week, in a Cuban hospital! In Qatar, maternity leave is too short – 50 days only! He was born on the day that the country was locked down due to Covid-19, so after maternity leave ended, I was able to work from home until he was 5 months old, at which time the nurseries opened again. The last 3 months were not easy, breastfeeding and diapers during and between virtual meetings, but it was a great opportunity to be with him for a longer period. The worst was, once again, not travelling to Portugal, as it was planned.

Another 12 months passed, following a routine between home, the nursery and work, without a social life due to Covid-19 and using the video-calls to communicate with family and friends. A huge support was provided by the company to all women to take care of their children during this pandemic – working from home when it's necessary and fewer hours a day at the office since the nurseries and schools were having reduced schedules or on-line classes. Finally, we went on a vacation to Portugal and I could hug my family and friends, after 25 months without seeing them!

## 15.8 UNDERSTANDING THE JOURNEY

I feel like a rich woman, my knowledge in a totally different area such as lubrication and reliability had a huge increase during the last 4years! My experience with different cultures has been amazing! And I still have a long list of concepts to learn. The community of experts in lubrication around the world is so big and the best thing is that we support each other on a daily basis by sharing our knowledge and experience. Every day, I'm learning something new about methodologies, technologies, different issues, possible root causes, etc.

I've also been the QA/QC coordinator of our department for 3 years! There are too many things to do under the different certifications to be better every day! Once again, I'm the only woman as the QA/QC coordinator within our group; all the other departments have a man as the QA/QC coordinator. I'm one of the internal auditors to perform internal audits within our group and external audits to our suppliers. Compared with the previous experience, as written above, it is the first time in the QA/QC department and the team is very-male dominated.

I had never considered any superior as a mentor until joining this company! Now, I have a mentor and it's a man! We are a small team in our department and we work closely, supporting each other every day in our different tasks. Our relationship is healthy, and even when we have hard/difficult days, everything runs smoothly. We used to say that we are a family outside our own countries.

It has been an amazing journey in STEM, and a great experience – personally and professionally. I was and continue to be a mentor for some of my previous co-workers. I hope to continue inspiring many more colleagues and especially women who go to STEM or to those who have been in STEM and for some reason are not anymore but would like to return. Here are some sayings that I would like to leave with you:

- Understand that each function/position can be replaced, but not a person! You are unique!
- Believe you can do it!
- You don't need to be the best, but do your best!
- Be your own competitor!
- Do not quit, never!
- Be confident, always!
- Learn something new every day!
- Never stop improving yourself!
- Share your knowledge and experience!
- Support all your co-workers!
- You don't need to have a mentor, but you can try to be one!
- Request help and support, at any time, whenever it's necessary!
- Make mistakes and learn with each one, we are humans not machines!

The long way taken through our life should be something we are proud of when we look back, including all ups and downs. I will leave you with one final quote, "*Strong Women are not simply born. They are made by the storms they walk through!*" – **Anonymous**

## ABOUT THE AUTHOR

**Vanda Franco** is a Portuguese female working in Qatar since 2017 within the oil & gas field. She is a chemist, with a master's degree in biology and marine resources and a postgraduation diploma in laboratory management. She has 18years of lab experience, and her theoretical training is complemented with the essential practical experience at the industrial and laboratory level in different areas in STEM, including marine chemistry, pharmaceutical, petrochemical and oil & gas industries. Constantly learning keeps her motivated both personally and professionally.

# 16 The Six Stages of Bringing Up Women in Engineering

*Kathy Nelson*
West Monroe, Minnesota, United States of America

## CONTENTS



Engineering, in general, is very male-dominated: this is obviously an understatement.

The first woman to be employed as an electrical engineer in the United States, Edith Clarke received a master's degree in electrical engineering from the Massachusetts Institute of Technology (MIT) in 1919.[1] Although Elizabeth Bragg was the first woman to receive a bachelor's degree in engineering (1876), she never worked as a professional engineer, but was a stay-at-home wife and mother.[2] In 1894, Julia Morgan received a degree in engineering, but went on to work as an architect, not as an engineer, designing more than 700 buildings. During the late 1800s and early 1900s, there were so few women entering engineering that they are listed by name in Wikipedia and other sources. One of the reasons for this is that most colleges didn't allow women into their engineering programs until the mid-20th century and some not until the 1970s.[3] Some women completed engineering coursework, but the colleges and universities often would not award them degrees simply due to their gender.

During World War II, there was a shortage of male engineers, so General Electric (GE) created an on-the-job training program for women with degrees in physics and mathematics.[4] Curtiss-Wright, an airplane manufacturer, hired more than 750 women who received training at seven academic institutions to work at their defense plants to help with the war effort.[5] Women have worked at the National Aeronautics and Space Administration (NASA) even before the Russians launched Sputnik in 1957, initiating the space race. Women also worked as "human computers" performing

DOI: 10.1201/9781003336495-19

highly complex mathematical calculations by hand at the Langley Research Center in Hampton, Virginia. Most of these female "human computers" were African American.[6] The 2017 movie *Hidden Figures* brought some of these untold stories before the public eye and may have inspired a whole new generation of women engineers.[7] Nevertheless, over 100 years after Edith Clarke took her degree, only 11.1% of electrical engineers are women. It took until 1972 for women to receive just 1% of the undergraduate degrees in engineering. (When I graduated from college in 1993 with a degree in electrical engineering from North Dakota State University, there were four women in my graduating class of 75 – just 5.3%.) In 2021, nearly 50 years later, women make up just 19% of all engineering degrees.

## 16.1 GROWING UP

Initially, I didn't set out to be an engineer. It wasn't a career I dreamed of as a little girl. I didn't even know what an engineer was when I was growing up, despite the fact that my father was an engineering technology professor who also wrote textbooks on circuits. At one time, he developed and taught a robotics correspondence course – virtual learning, but before computers were a fixture in homes and before there was an internet. He had always wanted one of his daughters (there were three of us girls, no boys) to be an engineer. When I think back on it now, I realize that he sometimes talked about wanting us to be engineers, but I had no idea what an engineer did, so I don't think these comments registered with me. We had no exposure to engineering in school when I was growing up. And so, although my parents were certainly fundamental in supporting their daughters' interests and strengths in science and mathematics, I wasn't really brought up specifically to be an engineer. In a way, I fell into my career.

Although there was no exposure to engineering offered in my K-12 education, there were programs to encourage children to think about math and science-oriented careers. When I was between second and third grade, I attended a weeklong day camp at the recommendation of my second-grade teacher, Mrs. Prom. During this camp, we each individually got to shadow a professional for the week. I don't remember what the focus of the camp was and, since this was the late 1970s, STEM was not yet a term people used. I also don't remember it being just girls at the camp. The camp was geared toward kids who were "gifted" in math, which makes me think it was probably STEM focused but not labeled in that way. What I do clearly remember is that I was assigned to a woman who was an architect. Now that I am older, I find this situation to be striking and probably unusual for the time. The fact that there was a camp in the 1970s where young kids shadowed professionals AND that the architect I shadowed was female seems remarkable! From that week onwards, I wanted to be an architect. That became my dream.

While I knew I wanted to be an architect, I assumed architecture involved math and science. My love for math grew throughout my schooling, and this I attribute to some key great teachers: my second-grade teacher, Mrs. Prom, referenced above; next my fifth-grade math teacher, Mr. Marsten; and my eleventh-grade teacher, Mr. Argent. Not only did these inspired people teach me math skills, but, through their own love for math and their passion for teaching, they also taught me to love math. They were "hard" teachers; they challenged me.

In twelfth grade, our state had its first year of post-secondary education option (PSEO), which offered eleventh and twelfth graders the opportunity to attend college classes in high school. I was able to take college calculus at the college where my father taught while I was still in high school. I loved it! Calculus was my favorite class and Dr. Legg was an amazing professor. I also took Physics – at the high school, not the college – and found a love for that subject, as well. The foundation for electrical engineering was being put in place. I just didn't realize it quite yet.

When it came time to apply for college, I applied only to colleges with architecture programs. I ended up attending NDSU in the College of Architecture my freshman year. And I was in for a rude awakening! While I expected architecture to use math and science – I actually needed high math and science scores on my ACT and SAT tests to get into the College of Architecture – my classes were mostly focused on artistic skills, such as sketching buildings on campus. One assignment was to build a structure out of sticks and stones and then sketch it, for example. I was no artist, so my first year of college was a struggle. I soon realized that architecture, while it had been my goal and my career passion for my early life, was not the right fit for me. But what was?

My soul-searching for a new major came between my freshman and sophomore years of college when I happened to spend the summer living with my sister and brother-in-law in Hermosa Beach, California. They lived a block off the beach, and I got a job working the lunch shift at a small sandwich shop on The Strand. I rollerbladed to work every day and after work would spend the rest of the day on the beach. It was a fabulous way to spend a summer in college. While I was there, my brother-in-law, an electrical engineer in the Air Force, was promoted to captain and I got to go to his promotion ceremony. We went to his office where they had his ceremony and then had donuts. I was thinking … math, science, and donuts … I could do this for a job. Shortly after that, I changed majors. Yes, my career choice may initially have been based on donuts.

After changing majors, I found my place. I loved my college classes – well most of them. And the power systems professor, Dr. Steuhm, made me and many of my classmates love power systems. Many of us went on to work as power systems engineers or at utilities because of him. Teachers had a huge impact on my career selection.

While I had been a gender minority in the College of Architecture, I was an outright rarity as a woman in my engineering classes. As I mentioned previously, my graduating class of 75 had just four women. I became good friends with many of the men in my classes, though, and being the rare woman in those classes never bothered me; they were my study buddies, my brothers, my friends. It wasn't until I started my first job as an engineer that I really felt like a minority.

Since those early years of starting out as a female engineer in a male-dominated field, I have found five "ups" to be important and it's also important to note that these are not specific to engineering, STEM, or even being female. They are just important skills that I have learned and continue to learn over the years.

## 16.2 SHOW UP

*When you show up for your job, bring all of you, as you are. Your authentic self. Your whole self.*

When I was a young engineer just starting my career 28 years ago, I went to work as a telecommunications engineer at a utility. I spent the first 6 months on the job out in the field with technicians. I was the first woman at my company to be out in the field. I worked with technicians as much, if not more than, engineers. Many of the technicians I worked with had 20 or 30 years of experience or more, and it was intimidating to work with them. I was just out of school, and I thought that I should know more than I did. For this reason, I tried to be "one of the guys." I wanted to fit in. I didn't want anyone to think that I didn't know what I was doing. What I didn't realize is that you're not expected to know everything when you start a job, especially an entry-level job just out of college.

While I talk about my dad having had a significant influence on my becoming an engineer, my mom had a college degree in Home Economics and I learned things from her, as well. I grew up sewing and baking and decorating cakes. I sewed clothes, including my wedding dress and some of the clothes I wore to work. I also baked and decorated cakes – fancy cakes. I baked cakes for birthdays and holidays for my family and friends. But, because I wanted to fit in and be "one of the guys," I never talked about these more "feminine" hobbies. I did as much as I could to blend in.

As I have gotten older, I regret not feeling like I could be my authentic self at work. I don't know why I wasn't confident enough in myself to be able to be both a woman and an engineer. Instead of celebrating who I was and what I brought as a female engineer, I tried to play down my gender. I am hopeful that as diversity and inclusion becomes more common in the workplace, other women will feel empowered to be their authentic selves and bring their whole selves to work.

These days, I try to ensure that I am my authentic self, or at least my authentic professional self, at work. Sometimes, I still worry about being my authentic self and speaking honestly, and I know authenticity is not always what everyone wants to hear. Nevertheless, I find that it is incredibly important and that usually when I am authentic, it brings value.

As I discuss later in this chapter, I started a podcast a little over a year ago. My podcast features conversations with women in STEM fields and I pride myself on having authentic conversations with the women I talk with. I sometimes worry that I don't sound smart enough, I don't know enough, I don't always use the best words, but I bring what is truly *me* to the table, my intentions are genuine, and I want to share women's stories. I am slowly realizing that I don't have to know everything about everything the women I interview talk about – how could I? Many of these women have PhDs in areas I never studied. I try to bring my curiosity and interest and desire to share their stories to my podcast. Listeners tell me that my interest and curiosity is shared by them, so in satisfying my eagerness to learn from these women, I believe I satisfy that of listeners, as well.

## 16.3 SIGN UP

About 15 years into my career, I was asked to join the board of directors of an industry trade association. I had been to conferences for this trade association for several years, but it was very male-dominated and not very diverse. It seemed like it might be a "good old boys" club. I would never have thought about joining its board of directors. But, due to some circumstances, I interviewed for a job across the country with a man who had just served as this trade association's Chairman of the Board. I ended up not taking the job, but a few months later he asked if I was interested in joining the board of directors. I agreed to meet with one of the board members at the next conference, although a bit reluctantly – this was not something I was particularly interested in or excited about. But I met some of the board members a few months later at the conference and decided to join. I was at a point in my career where I was looking for new growth opportunities. My job itself had gotten a little stagnant and I wasn't in a position where I could make changes. I knew I needed something else, and this opportunity presented itself.

While I say this organization seemed like a "good old boys club," I need to be clear that when I joined, I was one of only two women on a board of 80 members! As I look back on that fact now, it seems just crazy. But what I found when I joined was a very accepting and supportive group of peers who welcomed me with open arms. It was about this time that I found my voice and lost my fear of speaking up. Now, when I spoke up, it was in high profile meetings – meetings with Congress and federal agencies. I was honest about challenges we faced in our industry, and I was encouraged by my fellow board members to continue to speak up. Eventually, they encouraged me to pursue chairing the public policy division, which is an area I became quite passionate about, but again was not excited about initially. I point this out because I think it's important to be open to new things, even things that may seem intimidating or not exactly what you think you're looking for.

Getting involved with public policy and advocacy within this trade association gave me new perspectives – especially as an engineer. I had never made a connection between engineering, the work we do, public policy and regulations previously. I loved getting involved in public policy and advocacy. I went to meetings at federal agencies and Capitol Hill. Being from a small town in the Midwest, this was an amazing experience. I loved the challenges it provided, I loved looking at the bigger picture, and I loved advocating on behalf of other utilities and feeling like I was trying to make a difference in our industry.

Because of the work I did chairing the public policy division, I was encouraged to pursue leadership within the trade association. Initially, I was the Secretary/Treasurer, then Vice Chair, and finally elected as the first Chairwoman of a 70-year-old trade association. When I was officially elected at the membership meeting, I was given a standing ovation from my peers. It was one of the most incredible feelings I've experienced in my life.

The trade association I chaired was a global association; while most of its focus was in the US, it also had affiliations in Europe, South America, and Africa. As the Chairwoman, I traveled to and delivered keynote addresses at all their conferences. I traveled to Brazil, Portugal, and South Africa that year. I learned, with the help of a great communications coach,[8] how to be a better public speaker. This is an opportunity I never would have had while working at a small electric company in Minnesota. It was truly one of the best experiences of my career – and something I initially was skeptical and reluctant to "sign up" for.

## 16.4 SPEAK UP

I sometimes joke that I didn't find my voice until I was 35 and then I never shut up. I will make this comment for a couple reasons. One – I think it's important to find our voices early and use them. Girls are frequently told to be quiet and look pretty. We are not always valued for what we say especially when we are young, and many of us lose our voice, figuratively, at an early age. We may become "shy" and have a hard time speaking up in a group setting. Raising your hand in a large auditorium class can feel very daunting for some.

Research shows that girls' confidence tends to go down in their teenage years.[9] I think my own self-confidence probably declined in my early twenties, when I started my first engineering job. Understandably, in that context I didn't want to sound like I didn't know what I was talking about or didn't understand what others were talking about. I didn't realize I wasn't expected to know "everything" right out of college. The men I worked with had been at the company for 20 or 30 years. Of course, they knew more than I did. But that doesn't mean I didn't have good insight or perspective. I realize now that I could have and should have asked more questions.

There's a myriad of reasons we may be quiet and not use our voices as often as we should. But we all have important things to say, and we all have valuable and unique inputs. I remember a new female coworker who I was encouraging to present at a conference saying "I'm new. I don't have anything to talk about." I think it's normal to feel that way as a new hire. However, many people have been at their jobs and companies for a long time. They can appreciate and benefit from the new research, technology, and viewpoints that come out of universities. Being new in your field doesn't mean you don't have anything to contribute; *it means you have a different perspective, and possibly have new knowledge to share.*

Speaking up can also mean speaking up in meetings. Speaking up for other women or other minority groups if they're being interrupted at meetings. It can mean speaking up if you see injustices. It can mean speaking up for yourself and asking for that raise or promotion. It means being your own advocate and it may mean being an advocate for other women.

It can also mean speaking up if you are harassed or assaulted whether that's at work or outside of work. Unfortunately, this has happened to too many women. A few years ago, I was sitting with a group of women coworkers at happy hour. There were 12 of us – all women. I point this out because at that time in my career, the fact that I had 12 engineers or technical women to sit with was notable. Most of the time, I was the only woman, or at best one of a few, so being in a work setting with all women was unusual. We were discussing a serious harassment issue one of my coworkers had experienced. Every one of the twelve women at our table had either been harassed, assaulted, or raped. Now, think about that: all twelve of us. Most of us had never reported it and for the one woman who did, it hadn't gone well: neither with human resources within the company nor when she filed a criminal complaint. It's no wonder so many women don't speak up when these behaviors occur. This gets into a much more serious issue that is not what my story is about, or at least not this story. But, before moving on, I do want to say how important it is for us to listen to women when these issues come up. Because they do and they will. Hopefully the world is changing, but I have found change to be slow. This particular aspect of life is something I hope and pray will change faster so my daughters will never have this experience.

## 16.5 LISTEN UP

As I often say that I found my voice when I was 35 and haven't shut up since. While it's half joking, there is an element to that which is not so funny. While I found my voice and finally realized I had important things to say, as I moved into leadership roles, I have also found a corresponding need to be quiet and listen to others. In many instances this is even more important than using my own voice. And it can be incredibly hard for me to do. It turns out that I like to talk, and, as I get older, my thoughts evaporate faster and I'm always afraid if I don't speak, I'll forget what I was going to say. So, I need to remind myself to listen – and listen to understand, not to respond. I truly believe this is one of the most critical skills we can develop. If you talk to my co-workers, they may tell you I'm terrible at this; but if you talk to one of my podcast guests, they may tell you that I do okay at it. From this distinction, it seems clear that, for me, listening is contextual.

For many years after I found my voice, I only listened to respond. I wanted to get my input in. I wanted to be heard. And I interrupted others when they were speaking – which as women happens to us way too often. This skill is hard for me, and a constant area in which I'm seeking improvement. I now try to write down my thoughts on a piece of paper so I don't forget them and can provide my input at a more appropriate time. And I have found that not every thought I have needs to be shared.

I also want to help other women, particularly younger women, to speak up. Listening for those who are not speaking is also important. Listen for those who are silent. Call that person out at a meeting. Ask if they have a thought they want to share. They may have something to share, but feel intimidated or are shy. I draw on my experiences as a young engineer a lot, trying to remember how I felt and how I would have wanted someone to help me. The best leaders are those that listen more and speak less. I have a sticky note on my desk: "Be quick to listen, slow to speak, and slow to respond." That's my new goal.

## 16.6 GIVE A HAND UP

As I get older, I find it more and more important to bring women up along the way. To make their paths a little easier, to help them find their way, to provide support, and to create community. To make them feel a little less alone.

I've done this in a few different ways throughout my career. My most recent foray has been through my podcast "*Ordinarily Extraordinary – conversations with women in STEM*". This started out as a way to share women's stories when I found myself with some extra time on my hands after being furloughed during COVID. What I didn't expect was how much I would love my podcast, the amazing women I get to meet, and the feeling of building community. I found that so many women who wanted to feel like they weren't so isolated and alone in their jobs where they might be the only woman in their area.

My podcast has become my biggest passion and one of the greatest joys I've had during my career. Not only do I get to meet amazing women and share their stories with the world, but it has also ignited my own passion and joy for my career. I've met women who have jobs I didn't even know existed. I hope to spark young women's or girls' interests in these areas.

Another area of my podcast I have found to be incredibly important is education on what people do for jobs. While I had heard of engineers when I was growing up, I had no idea what engineers actually did. After many years of career and organizational experience as an electrical engineer, I now know that there are probably thousands of different jobs electrical engineers do. I want to let people know what we do for jobs so young people can have some understanding of the wide variety of jobs they might do when they make a career choice.

Earlier, I discussed the importance of listening skills. My podcast is where I get to really focus on developing my listening skills. My podcast is not about me, it's about the women whose stories I have the honor of sharing. I do talk about myself a bit, because I like my podcast to be a conversation and not an interview, but I try to sit back and mostly listen. There are so many women out there with stories to share. We all have a story to share.

Mentoring is one of the best ways to give a hand up. One of my favorite phrases I started using recently is that it takes a village to raise an engineer, or more generally, a woman in STEM. We all need our village, our support system. Mentors are a key part of that village. As we gain experience and knowledge throughout our careers, we can mentor younger women (or men) coming into our fields. Mentors provide advice, feedback, a sounding board, etc., which is vital for all people regardless of where they are in their career, but especially when they are starting out. The great part of being a mentor is that you can do it in different capacities. You can be a technical mentor, a leadership mentor, etc.

I have focused on being mentors to women, but recently, I have been thinking how important it is for women to also be mentors to men. Women mentoring men normalizes women's presence and leadership statuses in these male-dominated fields, so that young men become used to seeing women in leadership areas and getting advice from them.

## 16.7 SUNSET

I have loved my career as an engineer. While I somewhat fell into this as a career, it has been the right place for me and the best place for me. It's a bit strange to think of being in the sunset of my career. It seems like it just started. It seems like yesterday that I was just starting out. It's hard to believe I have almost 30 years of experience in my industry at this point. There are so many life lessons I've learned along the way. So many things I wish I'd have known earlier or done differently. At the same time, I have few regrets. Many of the things I wish I'd have done differently provided great learning opportunities and have been areas of growth.

Be brave. Be kind. Be yourself. Enjoy the ride.

## ABOUT THE AUTHOR

**Kathy Nelson** is an electrical engineer with 28 years of career experience as a telecommunications engineer specializing in wireless communications. In addition to presenting frequently at various engineering conferences and publishing in trade journals throughout her career, Ms. Nelson created and hosts the podcast "Ordinarily Extraordinary – Conversations with Women in STEM," where she interviews women in a wide variety of STEM careers about their personal career trajectories, challenges, and achievements. In 2017, Ms. Nelson was elected the first chairwoman of Utilities Technology Council (UTC), a 70-year-old global industry trade association representing utilities and their technology and telecommunications needs.

## NOTES

1 www.agnesscott.edu
2 www.swe.org
3 www.swe.org
4 Add reference for GE Program here
5 Add Curtis-Wright reference here
6 National Geographic
7 www.hiddenfigures.com
8 Anthony Vincent Bova, *EdgeWork Soft Skills*, www.edgeworksoftskills.com
9 https://www.theatlantic.com/family/archive/2018/09/puberty-girls-confidence/563804/

# 17 A Chemical Philosopher

## *Exploring the "A" in STEAM*

*Michael D. Holloway*
5th Order Industry, LLC, Texas, United States of America

## CONTENTS



According to the Encyclopedia Britannica,

> *"STEM, in full science, technology, engineering, and mathematics, field and curriculum centered on education in the disciplines of science, technology, engineering, and mathematics (STEM). The STEM acronym was introduced in 2001 by scientific administrators at the U.S. National Science Foundation (NSF). The organization previously used the acronym SMET when referring to the career fields in those disciplines or a curriculum that integrated knowledge and skills from those fields. In 2001, however, American biologist Judith Ramaley, then assistant director of education and human resources at NSF, rearranged the words to form the STEM acronym."*

## 17.1 DEFINING STEM

For me, STEM is a new term. Having been born almost 60 years ago, the term was only used to describe the thing that was between a flower and a root. I guess it's somewhat appropriate. The acronym 'STEM' (Science, Technology, Engineering, Math) provides a conduit for reality; if a flower is the be-all/end-all, the very essence of a plant's attempts at furthering its genetic goal of continuing then the stem is essential towards that goal. Without the stem, nutrients and water could not make it to the flower thus halting procreation. The stem provides support for the flower to be appealing for insects to pollinate it and for its seed to disperse. If the stem became compromised or worse, and if it didn't support the flower, then it would eventually cease to exist. That sure does sound like STEM. STEM is a support feature and a requirement for adaptation and advancement. Science is a "try", Technology is applying that "try that worked", Engineering is the method to get to technology, and Math is the foundation of the "try" being it an algorithm that is abstractly recognized or a trait that aligns the positive "try's" and eliminates the failed ones. Many other

DOI: 10.1201/9781003336495-20

creatures use STEM to manipulate their environment for increased success just not to the extent we do.

As children, we are born scientists, engineers, mathematicians, and we embrace all forms of technology for entertainment. We experiment and wonder. Math can be a fun game – almost magical. A LEGO set cultivates or engineering chops. When allowed the kitchen with all its raw materials and tools, it provides a fertile ground to experiment with food, condiment and herb spice combinations. When science is done right, children are enthralled and fascinated. Sadly, at some point in every child's life, science and math become difficult. Some continue mostly due to a born affinity, parental support or a great teacher, while most stop, and in some cases rebel against STEM because it has become a challenge. Higher concepts are not simple to grasp. Teachers won't always provide that sense of wonderment as children mature and enter higher education. STEM requires discipline. Discipline provides structure. Structure provides stability like a well-developed foundation made of granite and concrete. Granite requires time and work and energy to be useful. The result is magnificent but not at a cost. Concrete requires cement, water and aggregate with proper mixing and a mould as well as strict attention to the cure rate. If concrete cures too quickly, a runway exothermic reaction might take place and the result will be a severely cracked product lacking physical integrity. If the mix is improper, then the concrete will not set and have no integrity. The pursuit of STEM is no different – it is the very foundation for human advancement.

Like most children, I was fascinated by the world, and I liked science. I didn't love science. I would take apart things and build other things, yet my passion wasn't engineering either. My passion has always been the human condition. The mind. Back when you could subscribe to receive magazines at your home, my parents would get Psychology Today and the Smithsonian. In the second grade, I would read (or try) these magazines and kept that practice through college. I vividly remember discussing the works of Adler, Freud, Jung and Skinner with my mother at the kitchen table as she sipped coffee and smoked her menthol cigarettes. I was 8 at that time. She never graduated high school yet was one of the smartest individuals I have ever met. Later, in Junior High, we would have lively discussions on the works of Jean-Paul Sartre, Kierkegaard and Nietzsche. Joseph Campbell's Monomyth became my roadmap, Sun-Tzu was my mentor and Fritjof Capra (author of the Tao of Physics) became my credo.

## 17.2 THE PILLARS OF PERFORMANCE

My childhood had five pillars of performance. We were encouraged, make it required to 1) do well in school, 2) participate in sports all year, 3) go to Catholic mass, 4) take up a musical instrument, 5) serve your community. These pillars began when we were very young and stayed us through college and even beyond. I gravitated to the fourth pillar, musical instruments. I wasn't given a choice of which instrument to choose. If I was able to, I would have picked the trumpet. Miles Davis was about the coolest cat I ever saw! He made the trumpet sing. My mother liked Jazz as well but had other plans for me. She thought it best if I became a drummer. I fought it for a few minutes. It was useless. I was to begin my percussion instruction at the tender age of 7. So be it. As the years went on, I grew to love drumming. I had my own combo in fourth grade, playing popular ballads and hitting the talent show and convalesce home circuit. School band and orchestra, drum and bugle corps, percussion jam sessions, all help to develop my own style.

In high school, I was a drummer for a big band where I was the youngest by 10,000 years. It was more like 40, but it felt like the gap was much wider. I gravitated to various drummers such as Billy Cobham – the stick man for Miles Davis. Keith Moon of the Who, John Bonham of Led Zeppelin, Stew Copeland of the Police, Hal Blaine of Tajuana Brass and Terry Bozio of Frank Zappa, they all are considered great by all accounts of musicianship, but what made them influential to me was a combination of creativity, balance, coordination and syncopation. Each would play in such a way that sounded natural and easy until you attempted to duplicate a particular riff. They would produce original beats that had unconventional origins. I was to become the next great original drummer … until I wasn't.

By the time I was a junior in high school, I was destined to enter a conservatory until I had an existential reckoning – I didn't want to be a musician. Not that I didn't love to play, I did. I just didn't want to do it for a livelihood. This did not sit well with my parents. They had invested hard earned money for drum kits and lessons for over 10 years. I wanted to go to a college and become a doctor, specifically a neurologist and study the human brain. This still didn't sit well with them, but they eventually supported my position. I decided to major in chemistry. Arguably one of my worst subjects, yet I figured if I didn't get into med school or wanted to take a break before going, then I could always get a job in a lab. After all, how many chemists do you meet that are unemployed?

## 17.3 TO PURSUE OR NOT TO PURSUE?

Entering my third year at university, my advisor informed me that with a few extra classes, I would graduate with not only a BS in chemistry but also a BA in philosophy. I took many philosophy classes for enjoyment and enrichment without ever considering a degree in it. During a weekend visit to my parent's home for a good free meal and to get my laundry done, I informed them of my academic status. My mother developed a wide grin, my father looked up from his plate and said, "What the hell are ya gonna do with that? Sit on the beach and think?" to which I replied 'Well ya, but I plan on being pretty stoned while I'm at it!'. Normally many other fathers would have met that remark with a slap to the face, but mine would always break out into laughter with such comments. While the chemistry degree opened doors, the philosophy degree opened my mind.

I never went on to med school, instead I became a R&D chemist; then, I earned a graduate degree in polymer engineering. I have been accepted into several doctoral programs, yet found my passion in establishing competencies in others. I started my own company that is focused on developing the mind to understand, use and create. The science is the chassis, the technology is the suspension and wheels, the engineering is the drivetrain, and the math … well the math is the direction this vehicle is heading. Without the STEM, I would not be able to get this jalopy to go from point A to places unknown yet fascinating. Allow your young women to drive!

## ABOUT THE AUTHOR

**Michael D. Holloway** has over 35 years of experience in the industry. Having earned four university degrees, over a dozen professional certifications, authoring a dozen books and holding a patent, he started 5th Order Industry for the purpose of competency development and professional certification preparation. Holloway developed a learning management system based on the neuroscience cognitive pathway development that has been proven to be most effective at developing fast, sustaining competencies. He still drums but mainly to scare his dogs and frustrate his children.

# 18 Surviving Corporate Takeovers While Climbing the Ladder of Success

*Frances Christopher*
SGS, Texas, United States of America

**CONTENTS**



## 18.1 FROM PHOTOCOPYING TO UPPER MANAGEMENT

It all began as an accident. I was 18 and fresh out of high school. I was looking for full-time employment and stumbled across an ad for office help. After a rocky senior year filled with drama including a 6-week period where I ran away from home, it was a miracle I graduated from the all-girl Catholic school I attended. This job presented an opportunity to use my decent typing and filing skills and at $3.75 an hour I was hired!

It was 1982, the company name at the time was Cleveland Technical Center often referred to as CTC. CTC was one of the first commercial laboratory companies focused on testing in-service lubricants from industrial equipment. We routinely tested samples from engines, gearboxes, hydraulic systems, and other oil lubricated components. The working conditions were awful by today's standards, my boss was a tyrant and the employees hired were a mixed bag of characters willing to work for a minimum wage. At that time, my job was to make photocopies of test reports, mail them, and file them away for next time a sample was received. This was before computers, so everything was done manually.

Somehow, I stood out as being smarter than the average minimum wage employee and was offered a promotion to typist, filling in the cards used to report data then to billing, using the company's first computer … a Radio Shack with floppy disks having the size of a medium pizza. It's hard to believe that we could function as a business with so little technology.

Over the years, the company was bought and sold and changed names. I kept moving up in the organization and had mentors that encouraged me to go to school at

DOI: 10.1201/9781003336495-21

night and take on more challenges. I was a data diagnostician for 5 years, worked in customer service, and eventually became part of the management team working with some of the largest companies in the world. All the while, I was pursuing a degree at night, which I eventually earned after 9 long years.

## 18.2 GETTING THICK SKIN

What I experienced and learned working as a young female in a male-dominated business was that to succeed, I needed to have a very thick skin. I had incidents of sexual harassment, clients asking if there was a 'guy' they could speak to assuming I couldn't answer a technical question, and other little demeaning things that would sometimes take the wind out of my sails. I remember proudly working with a major OEM on a branded program launch and it was suggested (albeit jokingly) that we should have a ladder I could climb while wearing a wet t-shirt to draw people to our display booth. It's amazing what people could get away within the not-so-distant past!

I continued to work hard and impress my superiors and clients as an expert in the field of oil analysis. Incidents of unprofessionalism and disrespect are less and less frequent, but they still occur every now and then. When they happen now, it hurts more deeply but I still march on.

## 18.3 ACHIEVING LONGEVITY

I was with the legacy companies of Cleveland Technical Center for over 35 years, experiencing a period of incredible business growth, numerous changes in management and ownership, and a few unfortunate business developments along the way. The history of this business would make a great book or mini-series! My thick skin, resiliency, and ability to remain professional have been the key to my longevity in this business. I'm hopeful that future generations of women will not have the same struggles. There has been a significant amount of progress in changing corporate culture and more companies with zero tolerance for behaviour that demeans anyone.

My advice to women in the sciences is to let your knowledge and professionalism define who you are and how you present yourself to the world. Respect yourself and set an example for future generations of women. I've come a long way and hope the journey of others is easier and that women earn the accolades for their work and contributions they deserve.

## ABOUT THE AUTHOR

**Fran Christopher** has been in the oil condition monitoring business for over 40 years working with a wide range of companies to optimize their fluid analysis and equipment reliability programs. She has held a variety of positions, including data diagnostics, customer relations, key account management, sales and marketing, and operational planning and management. Fran holds a bachelor of science degree in business management and marketing and is a certified oil monitoring analyst through STLE. Fran is currently the U.S. Director, Oil Condition Monitoring for SGS, the largest testing, inspection, certification, and verification company in the world.

# 19 Embracing the Unknown, Overcoming the Past

*Kari Nathan*
Technology Exploration Career Center, East Lewisville ISD, Texas, United States of America

**CONTENTS**



Growing up in a small town is both a blessing and a curse. There were 54 students in my graduating class and I knew almost everything about everyone because most of us had spent our entire lives growing up together. My parents moved us to town so we could have more opportunities to be involved in the school and community. My parents are both hardworking, maybe to the point of missing out on some of what made me who I am. I think my drive for over delivering on my work expectations comes from their dedication to their employers.

Work was something I started doing at a young age, babysitting for my cousin's first born and also mowing the lawn for my uncle and aunt when my brother did not want to do it. I started filling vending machines with my brother when I was 15; at the plastic factory, my mother worked at for 35 years. Some of my skills, drive, and success are directly related to people needing help to get things done and my brother shirking his commitments. We are only 11 months apart and I am a people pleaser. He did not last filling vending machines, which meant I was doing it 7 days a week. It would not take too long, maybe a couple hours a day but it was like milking cows, people needed to eat so I needed to refill the machines daily. I did in fact milk cows on the farm that I was born at and my parents sold them when I was 5 years old.

At 5 years of age, I had one of many life-changing experiences. I was running a foot race in the school parking lot because the playground was too wet for play. My brother has told me the story, I do not remember. During the races, you line up and

DOI: 10.1201/9781003336495-22 

whoever wins is challenged to race the next person. I was racing a boy two grades above mine and I was winning. The boy's friend thought I should not win and clotheslined me. I hit my head and I was told I cried a long time, so long that the school eventually called my mom to come get me. My memory is gone, but I was told by my mother that they called after I cried for mom for over an hour.

With an older brother that I always worked to keep up with, my mother knew something was wrong because I never cried that much. Our first stop after picking me up from school was the local clinic. My only memory was breaking away from my mom's grasp and vomiting in the drinking fountain between the dentist office and the clinic. I went to the hospital in my small town, the one I was born in 5 short years earlier. I had a cracked skull and spent the night for observation. My parents tell me that my personality and coordination changed. I was Miss Majorette of Minnesota in baton twirling before the accident. The childhood memories I do have I feel like are not pure memories but stories others have told of my childhood. Maybe everyone has the family lures of their youth, but compared to my brother, I remember very little. I attribute that accident to the first change in my trajectory.

## 19.1 CHANGING MY TRAJECTORY

Working with my dad as we remodeled our home was something I always did and I am not sure why I was always willing to forgo childhood pursuits to work, but my guess was it was the main time I got to spend with my dad. I learned all kinds of skills like dry walling, gardening, framing, electrical, demolition, hanging wallpaper, painting, building a deck, and the list goes on. My father admits he was not the best teacher in the earlier years because it was easier to do than teach, but I was an observant and wanted to impress him so I would do things like paint the living room while they were at work. I would wash dishes so that my parents would have more time, did my own laundry, mowed the lawn, raked leaves, and learned how to do most every household chore. My grandma lived only two blocks away in the town, so I was helping her whenever she needed it, cooking, cleaning, holiday decorating, whatever she needed. I never knew she was really teaching me the best ways to clean and cook. She was always one of my role models.

My grandfather died in a tragic car accident when my dad was 24, I never knew him but I attribute some of my strength in self and feminism to the fact that my grandmother was 48 when she lost her husband and he was also gone during the war. She was able to keep everyone healthy, happy, and always had extra to give. She worked at a hardware store in my hometown and I always thought I would work there when I was old enough. I felt like I was training for a future job when I would run to the store to get something for my dad's project at home. I remember watching one of the other employees fixing a screen window. The hardware store was one of the places I liked to observe other than being the library.

Being a kid in town with a bike at that time gave us freedom to explore in a safe town. My father had lived in the town his entire life so most people in town knew us. This knowledge not only made it safe for us to be independent but also reminded us that parents would find out if something went wrong. Sports was my favorite pastime. Our neighborhood had four to five kids who were around my age. We played basketball, t-ball, football, and night tag. We all went to the summer recreation program at the nearby ball fields. I know I got moved from t-ball to the next level up ponytails since I was good for my age. It really might have just been that they were short players, but I thought they thought I was good so I wanted to be good.

Once we were in town, I always played sports outside of the organized teams with my neighbors. They were all boys and I never noticed that there was much of a difference. I always equated sports with my neighbors and school friends with my organized girls' teams. Because one of the neighbors' dad wanted his boys to be good, he would sometimes play with us and coach us in mostly basketball. I always got to play and I think playing with the guys was better than playing on my girls' teams because the attitude with the guys about sports ended on the court and they would not hold a grudge for what happened on the court. I loved in gym class when teams were being picked, before crushes, I was the first girl picked for most sports. Hat tricks in floor hockey, catching the hard throws in dodgeball. I was never afraid to get hurt because growing up with my brother and neighbors made me strong and not emotional about injuries.

My friends were all better off financially than we were so it made friendships strained at times. I never wanted friends to come to my house or have sleepovers. My best friend growing up was Jaime and she had everything. She was great because she also liked sports and competitions. I know I must have been to her house monthly and never did I invite her to stay at mine. Her parents would drag me along to movies, skiing, bowling, or whatever Jaime would invite me too. My parents always had enough to take care of us and they worked hard to allow me to go skiing and bowling whatever I wanted to do. Because I started "working" at a young age, I could pay my way, but I never thought about having her over. I do not know if I was embarrassed because our house was constantly being remodeled or if I was just insecure about our families' different quality of life.

My friends never asked to stay over either so I am not sure if parents were worried about our house because they never met my dad. He worked hard and rarely made it to my games, concerts, etc. He is a quiet man so that might have made him intimidating too. I have small memories of wearing clothes from yard sales that kids at school would be like, "I used to have that shirt". I wore cut off jean shorts to softball one time because they were clean and a girl made fun of them because you cannot play sports in them. I was envious of other kids put-together clothes and hair at times, but mostly I would just live my life. That is another reason I spent more time with the boys. They did not care what I was wearing. All of these factors contributed to my competitiveness and comfort with the guys around me.

## 19.2 GETTING LOGIC ON MY SIDE

Planning for sophomore year of high school, there was a small group of students who wanted to take two math classes at the same time so we could move on to college math in high school. I was the only girl and I was never worried about it. Over the years of sports, the boys had a respect for me since I could hold my own with them in the classroom or on the courts. We took two math classes every year after that in the high school. AP calculus in the high school did not go well my senior year, I was the only girl in class and it was taught over interactive-TV with the teacher in the next town 8 miles away. Only six students fit on the camera, so I was the one who was off. The teacher had never taught over interactive-TV and calculus at the high school level. It was difficult and I was not self-disciplined with my education. I was the one who would leave class every day to buy cookies from the cafeteria during class time before lunch. That did not help my skills in calculus. I still got good grades, but I did not test out of any college calculus. Because my parents were not very involved in my schooling, I made some foolish mistakes that I think if they had more experience or involvement in my education they would have made me finish some of the things I started.

I was mad at the volleyball coach after my junior year because I heard her talking about keeping a freshman on the team and in my mind that meant I would lose my spot on the team. Once I realized that it was a mistake to quit, it was too late. Basketball was better, I knew I was one of the best on defense and was excited because we only lost two players to graduation. Our coaches changed almost yearly and the head coach my senior year I thought was playing favorites. I asked one of my teammates if I was crazy and she said she thought I was not getting enough playing time and recommended to talk to the coaches. I first collected some data since I am very pragmatic. I had the same points as another girl in the same position, more rebounds and considerably less playing time.

I spoke to the coach before the game, I had logic on my side, I was ready for him to apologize and let me play more. I kept my composure and explained the numbers. An adult nearby got his attention so he cut me off and said he would think about it and talk to me about it later. I was nervous, but I had logic on my side and my point guard agreed with me. The coach came into the locker room before the game and got everyone's attention and called me out in front of the team. "Kari thinks she is not getting enough playing time and thinks she is better than her teammates, I will give her more playing time and we will see how it goes." Another turning point, I was the high scorer and had a triple double. That was my last basketball game in the high school. I had an ankle injury and was putting off the repair surgery. I hated how he belittled me in front of my team and I scheduled the surgery and had it before our next scheduled game. I thought again he would apologize and tell me I was right but the truth was he never gave me another thought.

Being invisible or unremarkable in a small town is very lonely. All through my teenage years, I felt like I was on the outside looking in on the popular kids. Surely, I was not the only one in my shoes, but I felt like in my grade that I could not overcome one last obstacle. I never knew what it was but I wanted to be included and never knew why I was just outside. Blaming my brother, my lack of money, my lack of interest in the popular boys as a boyfriend, my parent's careers, whatever I could think of, but I never could cross the boardline from average to popular. My high school career ended in the most heart jerking manner, but I might have been the only one to notice … 54 in the graduating class, there was a slideshow of pictures of the class members at different ages and doing different things. Everyone was in it, almost everyone. I am not sure if I was the only one missed but it felt like it. Not one picture. I had seen that before in the newspaper with sports, but for some reason, the fact that the yearbook club forgot me, it hurt.

## 19.3 I DID NOT KNOW WHAT I DID NOT KNOW

College felt like it would be better, I applied to three places and got in all three. In my junior year, I took the military exam of basic knowledge because it was the best way to show your strengths. I was in the top 2% of people taking the test in mechanical aptitude, so engineering was on the top of the career lists. For lack of a better idea, I focused on getting ready for engineering school and applied to the University of Minnesota Institute of Technology, Iowa State University, and University of Wisconsin River Falls as a back-up economic starter school. I got in all three. The truth was that my parents made too much money for me to get a full ride scholarship or loans and not enough to really be able to help pay for it. Pragmatic Kari thought I will go to River Falls for my first 2 years and transfer into U of M for the last two to save money. Again, I was the only girl in most of my classes in my major, but I was used to it. I had a hard first semester with a professor not turning in grades for computer programming and I did not go to my final in Calc-based physics because I thought my grade was too low. I did not know what I did not know…

In the college, average grades differ by program and professor. My low grades were average, I would have passed if I had taken the final. I got put on academic probation, and a letter went home to my parents. My dad said, maybe you are not cut out for college. It hurt, he should have been my biggest cheerleader, but since my parents did not have any first-hand experience with college, they did not know this was normal. I got a 'B' in computer programming when my professor finally turned in grades. Three full years I went to UWRF and graduated in August with my BS in agricultural engineering technology. After learning, I could graduate in 3 years the U of M seemed like a waste of money. I thought all degrees are the same, like a driver's license. I did not know what I did not know … I was wrong, if you do not have a family that can guide you in college, find a mentor. I am not sorry I took the path I did, I am sorry that I did not have teachers in my hometown advocating for me. If my degree would have been from a more prestigious school, I would have had a jump start at my career. I might have also taken a different path that would help me find passion in my work earlier.

## 19.4 BEING OPEN TO MISTAKES

Fear of the unknown has never been an issue because I always work at something until I succeed. Missing a test for my engines class, the professor sent me to her office to take it. I went into the wrong office since I did not know one of the doors in her third office description was a closet. As I was taking my test in the wrong office, I met the head of the department. He could not look me in the eyes but told me I was in the wrong office. I could have fled in haste, he seemed unhappy with me. Luckily, my favorite engineering professor walked in as he was kicking me out and told the department head that I would be a good candidate for his composting research project. This mistake got me my first internship. Being open to making mistakes has led me to many opportunities. Another professor told me to apply for a summer internship with New Holland NA, the recommender warned all the local opportunities were full for the upcoming summer, but they thought I could get a jump start on the next summer. I was invited to an interview and asked about milking cows and driving tractors. The interviewer said "if you have milked cows you know how to work." It happened that the farm that I milked cows on in my high school had a New Holland BiDirectional tractor. They called me and asked if I was willing to move to a different state for the summer for an internship. They would make a special spot for me in Indianapolis.

As an intern, I worked at a small commercial business unit that managed dealerships in Indiana and Michigan. I was sent to train for a week in Pennsylvania where I learned about the history of the company and operated most of the equipment they sold. In Indiana, I worked with a staff of older employees, Carl was the only younger guy but a decade older than me. The rest of the staff could retire within the next 5 years or less. I was better at computers than many of them and they really did a great job making the experience great for me. They needed my help with the computers and they gave me great work experiences. I got to "work" a Nascar race and invited a friend to join me there. Perdue Haydays and Michigan State farm show were events I got to work. I had limited product knowledge so at one they had me move a tractor pulling a baler up in the demonstration rotation. The one I was on first was one I had used before with a baler I test drove in PA, the second one was a tractor that the PTO gauge was not working and I had never used it or the baler before.

Everything was fine until they stopped the rotations with presenters and had everyone keep baling in a circle. I was on the broken tractor with the baler I had never used before. A salesman from the company Chuck walked up to the tractor to show me how to turn on the PTO and told me to give it hell. I nailed it, no issues, no plugs, good bales. Everyone was impressed by my success on our team because they knew the secret that the PTO gauge was broken so I had to guess when it was at full power by sound alone.

Customers including the Amish came to talk near me, not to me, to the sales team that the equipment must be easy to use if that little girl could do it. (I was 20 years old with a babyface, it was a fair assessment). The next show in Michigan they had me drive the bidirectional tractor the first day since I knew what I was doing, so I had to go to the baling demo field collect the bales and put them on bale wrappers and tubing machines. This again was impressive since I was a "little girl" and doing an expert job and keeping the other short-line manufactures equipment running. They keep me on the tractor for the entire show.

Fast forward to January and my interview with New Holland for full-time employment after graduation. The first piece of advice is to apply early, if they want you, they will wait for you. I interviewed in January fully knowing I would not graduate until August. My competition was still the students graduating in May. I knew from my internship that most employees came from Penn State, The Ohio State, and Cornell. I was a female in a male field that made me abnormal at the time, especially since I had experience. I did not know what I did not know … I was flown to New Holland Pennsylvania over my winter break. I had no idea what to do to prepare myself. My friend and RA at school helped me make a plan. She was studying agricultural business and communications. She said long skirt-suit past the knees (important fact) and subtle colors. I was in a pinch, bought the right suit, and found a long black leather jacket on clearance that was 3 or 4 sizes too big, but I could afford it and it was better than my ski coat. I printed out several copies of my resume, bought a black leather portfolio, and thought I was ready. Ready or not, I was in New Holland and someone drove me from the hotel to the headquarters.

I do not remember those details, but it went smoothly. The first interview was a round table interview with eight people interviewing me. It was unsettling to walk into a room with that many people judging you, and judging you critically with a fair reason. If I got the job, I would represent the company in service of machinery, and at that time, "women" did not know that kind of stuff. I am not saying that the company was sexist, I do not think that at all but they did know their customers, which were older generation farmers. It was like a war zone, questions were being shot at me from all directions. Which class was your favorite, least favorite, why. Five-year plans, strengths, weakness, ability to travel, goals, and internship questions. I was very unsettled after that interview, but the rest of the day should be easier, a guy was taking me to lunch and a tour of the headquarters. If you are unfamiliar with farming and farm-related business, it should be stated as fact that I should have known the person taking me to lunch would be driving a truck. Not just a truck but a big truck. I was in college with all kinds of farm kids, our college was jokingly called Moo U or Silo Tech. I should have known. I was unprepared.

Geometry and muscle let me down that day. I did not slip on Ice, I am a Minnesotan, I did not spill food, when I was getting into the truck to go to lunch my skirt ripped at the seam in the middle of the back. I heard it, I did not know what to do or how bad it could be. At the restaurant, I excused myself to the restroom when he was distracted, in case it was bad. It was only about an inch higher than when I bought it. I was in good shape until I had to climb back into the big truck to go back to the headquarters. It was like the rip was the only thing I could hear and worse was the cold draft I was now feeling. What would I do? I walked a respectful halfpace behind him for the tour, I was cautious not to bend over or walk in front of him on the stairs. I heard a few more stitches letting loose on the stairs. If I pulled it up over my knees, it would not rip but, I mean "butt" I might put on a show. I do not remember the tour, only the sound my skirt was making as one by one the stitches let me down. I did get the job. I had less than an inch of modesty left by the time I got back to my hotel to change. That was the first and last time I wore that suit. Advice, always try your interview outfit out for flexibility and modesty.

## 19.5 GETTING INTO AGRICULTURE

Agriculture today is still predominantly for white males. It was more than two decades ago when I was hired as a service representative. I got to my training in PA before moving to Albany, NY, where my job is located. I was working at a different office to train me a little because they were short staffed in Albany and some of the people worked from the field not from the CBU office (commercial business unit). I flew into PA by myself, it was the right way to start a job but I of course was unsure of what would be happening. Luckily, I had and still have some good friends from my major in college that worked for the same company. None had worked in the role I was given. The office I trained at had two women working there, but they were secretaries not in service or sales even though they did a lot of sales work. They told me

I was filling a sales opening position, I said I was hired for service. This worried me, but the training was on things like expense accounts, company car, HR paperwork, computer systems, and office flow.

After about 2 weeks in that office, they said I was ready to drive to Albany in my company car, a jelly bean blue Ford Taurus. The lady I was replacing had ordered, it but she did not get a car in her new position so it was mine. It was the newest car I had ever driven. I thought the color was a poor choice for a lady in a male-dominated field, but there was nothing I could do about it. I was greeted at the office by Bill, a tall man wearing too-short running shorts, navy blue with white trim, long striped top knee high socks, shoes that looked like they were bought in the 70s, and a nondescript t-shirt. My first impression was underwhelming. The office had a small staff of 5 in sales and 3 in service. Bill informed me that I too was in sales, I again said service. He said he would make a phone call and go meet everyone. Tom, the youngest in service, said if they say sales take it. It is a better position, service is a dead end. I was in fact in sales.

There was a transition in the office structure and everyone was told they had to work from the office when not on sales calls. So, in a short time, the Manager Dick retired, John took over, and Bill left for a promotion. During training, Dick wanted me to ride with each salesman for a week to get a feel for the role and for different styles of handling the job. I learned that Bill bought everything on his personal credit card instead of the company so he could get the point, he also shopped for beanie babies for his daughter's collection, she lived with her mom. At Apple equipment, Bill neglected to introduce me and the owner John said "Are you here to keep your dad honest." I felt like it was more of a dig on Bill than me but I did not like being the kid. Bill did not help with the situation because he did not want me to talk, just observe. The younger sales team members were better. Nate was the closest in age and he is Black so he had the same problem of proving himself to the old white men we had to visit.

John refused to take me out on sales calls. I never knew why and when Dick retired and John became my new boss. As any new employee, I had things to learn but because of my internship, I knew the computer system. I had mentors in a different office that were not threatened by my presence. They would email or call and check on me occasionally. I started making dealer visits and they gave me all the tasks they hated, hazing the new person I am sure. It was all new to me, including the North East. I was put in charge of Government Bid and Fleet, because Chad took another job and the old guys did not want to do it. The paperwork was cumbersome, but the physical lease equipment inspection annually or for trade-in meant hunting through the largest nursery on the east coast for the equipment and checking hours. There was never a feeling of appreciation or respect by John. I remember he asked me about golfing. He said if I did not go to the dealer for golf outing, I could stay and cover the phones. I did not tell him I had never golfed but told him I was not very good.

I went to the golf outing and luckily my former athletics made me ok for my first time golfing using clubs I bought at Sears days before. One of my teammates for the day was a very good golfer and taught me a lot that day. He loved that I could hit the long balls and ended up using more of the T-shots than required. I never told John I had never golfed before and the dealership people I was paired with for the event never told him either. I was in New York for about a year when my long distance boyfriend graduated from River Falls with the same degree. The day after I got back from the New Holland interview, before we were dating, he asked me about the interview and I shared the details of the skirt conundrum. He laughed and shared a story about an interview he had for a summer internship, his pants zipper was down for a long time until someone at the interview told him. Side note on engineering, if you find time to date there are usually more guys than ladies, I did get the best one from my department as my husband of over 20 years. He had job offers at several places after college, but the best choice for his career based on his specialized skills was in Omaha, Nebraska. I fully supported him, but I knew we would need to live closer if we wanted a future together.

The money was good, but I was not feeling valued at work, so I spoke to the regional assistant manager about working towards a master's degree in adult education and training. The biggest obstacle I told him was all the schools that had the program were in the Midwest, could I transfer? He recommended I should transfer to the Ames, Iowa office. They were short staffed and I had won the company excellence award all 4 quarters in Albany. (Gold watch with company logo on face and a diamond each quarter after) This was an easy choice since I did not feel appreciated by my direct supervisor.

Ames was 2.5 hours from Omaha and 5 hours from my hometown. I was excited to move and hoped I would be appreciated. The new boss Keith was from Canada and he was a great guy. He loved all the skills I had learned in Albany and was willing to teach in Ames. The sales department was young, and I think other than two employees, I had been with the company the longest. Another trajectory change came when one of the dealer principals started "flirting" or harassing me. He was in his 70s and I was 22 at most. I was not prepared for this from college or previous work. Working in a hardware store through college I did have my fair share of weird experiences. Ilearned from one customer in the middle of winter in Wisconsin that women have an extra layer of fat, so I did not need a coat to go outside and cut him 150 ft of wire.

My experience with home improvements left me confident and I was always in a store with lots of people. The dealer principal was at a very old small dealership, he was on an improvement plan, which is business speak for not a good dealer. He was not selling equipment, I think it was 4 years since he last sold a tractor. I was taxed to visit him quarterly to get him selling again. We had dealer standards inspections, we would go through a 100–200 point check list for sales and service. We had to do every dealer in a set timeframe and it was a team effort, one sales and one service person. When we arrived and the dealership things were going ok, he would make some small comments, but I was one of the guys, I was trying not to let it bother me (many shops still have calendars with underdressed women on the walls). This guy had a naked hula lady clock on his desk, stacks of yellowing papers, and used cans of "new" paint for sale. His dealership was not the place any woman would want to go, I was getting paid so I did.

When we were leaving, this Creep had the audacity to come over and hug me goodbye. I was in shock, I froze, I was a sculpture of a person. My mind left my body behind and I was left dumbfounded. The service representative I was with got him off me by saying something and I got the hell out of there. The company had me document all his lewd comments over the phone and any interactions. He would only call to talk to me. My boss was supportive and told me I never needed to speak to him, but if I could put enough information together, the corporate lawyers would terminate his contract. It was a few months of uncomfortable phone calls including an invitation to his grandson's college graduation in Kansas with him to get his store closed.

After that, the dealers were worried about having me visit their dealership since I closed the creeps. I had one tell my boss that he did not want me to support his dealership because he was afraid he would swear in front of me. This dealer I met at the Clay County Fair in Iowa. It is the largest county fair and bigger than many state fairs. After 12–14 hours a day working out at the display booth, he asked when I would start calling on his dealership. He realized I was good at what we did and was very good at the logistics of the equipment. I just laughed and told him to ask my boss. Things went well in Iowa, we had big equipment training at our office since it had a big showroom. My old boss John was in town for the training and he took me aside the first break and apologized to me about being a crappy boss. He said that he never knew how good I was because he never had to train new employees. He said that the two new guys who started after me had struggled and needed hand holding. A complement from him was rare, it was too little too late for me to ever like him as a colleague.

## 19.6 BENEATH THE SURFACE

Experiences like these have shaped me into a teacher and a person who always thinks there is more to people than the surface. I was married in August 2001 and honeymooned in Alaska for almost 2 weeks. I liked what I was doing and I did not apply for new jobs before leaving the company for marriage and Omaha. I did not know what I did not know … My husband was traveling testing equipment in the field and I was in Omaha unpacking and looking for work. He called me on 9/11/01 and woke me in a state of odd excitability. He said turn on the TV, I said, "What, Why?" he said just turn it on. I was floored, I saw the planes crashing into the World Trade Center over and over for the next few days. This event changed America and my career path for good. Work was hard to find, the economy and country was in a state of change. I got a job in October working for a telemarketing company selling lists of names to the market by demographics. It was one of the worst places I have ever worked. Total toxic environment. People stole your sales, the owner was notorious for firing people during meetings or giving them a weird Eagle coin of no value. Someone took a crap in someone else's trash can. Toxic, I got out of there as fast as I could find a reputable place to work.

With my sales experience, I was able to get a job selling credit card processing services at one of the largest and first processing companies in the US. Again, high pressure telemarketing sales was stressful. Two hundred cold calls a day, a boss that would walk by your desk and log you off your phone (timeclock) if you were not at your desk. You could be faxing papers to a customer or using the bathroom, he did not care. I ended up with migraines for the first time in my life and depression. I was good at math and sold with logic, but he would always tell me I am not doing enough. When there was an opportunity to switch to personal banking, I took it. I worked at a branch that had the reputation of being high sales but a tough boss. The lady in-charge told me to wear more skirts, maybe I should dye my hair blonder, and wear more make-up. She had skirt suits and helmet hair. The best day of work was the day a nicer middle-aged man came to see me about his daughter's account, she kept over-drafting it at college. He said to me after about 45 minutes working together to solve her problems, "Don't take this the wrong way, what would you rather be doing for a living?" Shop teacher was my response and he laughed.

This was the day my direction changed and I feel like I have found a big reason to work hard and change the future. I said, "I know, a lady shop teacher is not normal", he was of that generation. He laughed again and said "No it's not that, that's what I do." Three phone calls, and 2 weeks later, I was hired to teach industrial technology. The transition to a teaching program allowed me to get certified to teach while teaching. I continued on to get my master's degree in curriculum and instruction in industrial technology since it was only a few more classes. I learned from a young age to give people more than they expect as an employee. My principals were good and my mentor was great. She is still one of my best friends 18 years later. The students were middle school age and they taught me more than any class ever has. Middle school is a hard time in a student's life and making connections with teachers might be all it takes for a kid to change their path or find their way. Four years into my teaching career and there was an opening for an engineering teacher at the Magnet High School in the same school district. I applied since I was the only teacher with an engineering degree in the district as far as I knew. The school was in a tough part of the town, but the students were a mix of socioeconomics due to the magnet program's recruitment. The thought of not getting the job did not cross my mind. I had recently finished training for STEM education with NASA in Houston. Robotics was one of the new curriculum pushes and I had just learned about how to teach it from NASA.

Overlooked for an interview, that is what happened. I asked HR why I didn't even get an interview. She said that the decision was made. I asked if the new teacher had an engineering degree, she said no. This made me bitter. I thought that in 2007 the best qualified would be offered or at least interviewed for the position. Timing is everything. My husband was offered a chance to go work overseas for his company and I pushed him to say yes.

Let me say personal issues are always hard to talk about and help us make choices related to goals. I have always thought I would be a mother, it was something that I felt like school told us don't have sex, it is easy to get pregnant. The problem with education is that it does not tell you that this is true as young girls/women but you reach a point that it is less likely to happen. My husband and I waited for 2 years to start trying for kids after marriage. We invested tens of thousands of dollars for fertility tests and treatments. It did not happen. We were moving into our second home that we built as general contractors when I was 29 and he was about to turn 30. It was after his birthday that I discovered that I was pregnant without any treatments.

Nervous was an understatement, we said we would not tell anyone until the 12-week checkup. At the 12-week appointment, the ultrasound tech left the room, the doctor came in, and sent me to a specialty clinic. That clinic sent the results to my doctor and there was no heartbeat. I had to have surgery to remove the "growth" and the doctor said, "You got pregnant once it will happen again." Wrong! My husband took the job in Germany and since IVF was less than $5000 per attempt versus the $20,000 in the USA. We tried multiple times with crazy hormones and weight was all I gained. Now I am 45 still have no kids, tried adoption for 10 years after getting back from Europe and never was matched. This is a very upsetting problem. As a school teacher and an engineer, we hoped we would have been picked to adopt. Here, I sit writing in our almost a million dollar home with our expensive cars and no kids. Money cannot buy everything.

## 19.7 THE UNKNOWN OF TEACHING MIDDLE SCHOOL STUDENTS

My husband was due to transfer back to Omaha from Europe in September after his 2 years in Germany and France. Besides teaching a little English, I did not work in Europe. I did finish my school administration certification. Getting restless, I applied at my old school district. Again, no call from HR. Annual reviews were all great so I was taking it as a sign to look to industry for a new career. I got a call from the head of career education at my former school district when I was in Minnesota for a friend's wedding. He wanted me to take the job at Nathan Hale Middle. The teacher hired quit before school started so they had been a couple of weeks with a substitute teacher. HR finally called and asked if I would interview? I interviewed a couple days later and took the job. Husband and I would be apart for a month, but we had endured greater hardships than separation. HR told me I could wait and start at the end of the quarter or December if I had to get things in order. I told them that was not good for classroom management and started ASAP. My husband's job got extended in France a few times and I was not able to move home until 6 months later. He was home for Thanksgiving and Christmas. It was not easy with the time difference, but I needed to stay busy and stop worrying about kids.

Teaching has given me the opportunity to work with thousands of students. It has been very rewarding and challenging. I have learned about hardship, students who slept at school because they still had nightmares from the war in their home country where they were wielding an AK47 and fighting for their lives. The students who are ruff around the edges are very challenging but very rewarding. Middle-school students are resilient, but the tragedies are hard to hear. Kids who get removed from their homes because in their family "It is ok to have sex with family." Families that are living out of their cars, students who need to shower at school because they have no running water, but the school locker room is filled with obsolete computers. Parents coming drunk to a middle-school boys' basketball game and berating their kid at halftime because they are not hustling enough. Kids in foster care wear ill-fitting and dirty clothes, while the foster parents' natural kid is wearing new Jordans every month. Same kid in the seventh grade gets pregnant with an 18-year-old down the street and the foster parent does not press charges and keeps the foster kid and the new baby in the home and let them live in the unfinished basement so that they can have their own space. Heart-breaking when these things become known, as a teacher you learn to limit the information you learn about students because it will break your heart. I have called CPS at least four times.

The high school I taught construction classes at was the best experience I have ever had as a teacher. My favorite success story was a boy who lived with grandpa and step-grandma, Zach. As a 10th grader, he was a mess. Laying on work benches, talking trash about other students and his neighborhood. He was fronting hard so that he would not be a target. His dad was in prison for stealing copper, his mom was a drug user and out of the picture unless she needed something. His grandfather was wheelchair bound because he lost a foot from diabetes. His uncle is 1 year older than him and his brother is 1 year younger. Over the years, I learned a lot about his family. He was the stable and responsible one in his family.

He would go rescue his half-sister if his drug addict mom would call tweaking and he would go pick his sister up so that his mom would not drive. He would protect his brother from drugs, alcohol, and gangs. His uncle dropped out of school his senior year and was a freeloader according to Zach. He had jobs since he was 14 and was too proud to ask or take help. He was good with tools and easy to encourage to work hard learning trade skills. I was able to provide the support he needed and encouragement to do well in school. He was the first person to graduate high school in his family for generations. His senior year I helped him get hired by the company remodeling the school. He was very good at the work and did everything the employer wanted and more. The Sherwood Foundation (funded by Warren Buffett) bought him work boots and a jacket for work. With help from a few teachers, he was able to apply and get a full ride college scholarship from the same organization. He is currently working to become an electrician but did not stay in college. Another adult convinced him he should not go to a small state school and go to a large university. His first math class had 300 students so he did not go. I was worried I had failed him because I had moved to Texas.

Talya was a student lawyer in seventh and eighth grade. She was the student who would defend her peers or remind the teacher that they were not being fair. Always quick to raise her voice and advocate in a negative way in class. I thought she hated me as a teacher. I had her again as an eleventh grader, she was a delight in class. She was the only girl in her construction class and she would help keep the boys in class in line. It was the idea that every day was a new day in middle school and I never held a grudge with an awkward middle schooler. She worked hard in class and she took another course with me as a senior. Her attitude had changed, she was still an advocate but she learned how to control her emotions and use her intelligence to improve situations. The unknown of teaching middle school students in a big school district is that you do not see the students' improvements because they go to different high schools and move on. The rare few I had as both middle-school students and high school students was very enlightening, the growth was amazing. Taking some students on a college tour I ran into a middle-school screwball and he ran up and hugged me. I had not seen him for 3 or 4 years, but he remembered me and was so proud of the successes he had at college.

Teachers who can make genuine connections with students can have a huge impact on students. Getting text messages from former students who still use me as a reference, the new jobs, weddings, and homes are things that students let me know about. I have had late night "drunk" texts that they did not know why I put up with all their crap as a student. Not only updates on their careers but also their families.

One has been keeping me posted on his dad's recovery from Covid, hospitalization, Acoma, intubation, recovery after being in a hospital bed for 3 months. They become family when you can teach them for 3 years and watch them grow from little hot messes to fully functioning adults. They fill the hole with the lack of my own kids but they also remind me I do not have kids.

The connection between teaching and having our own children was the concept of having a schedule to be able to be more involved in our kids' lives … No success and it has left me behind in my professional career of engineering. I worry that the years I have spent teaching have left me behind in industry. Insecurity in my knowledge base has left me tied to my teaching career that constantly reminds me I cannot have kids. I know I am making a difference in many students' lives, it does not make it easier to know kids will never be in my future. My recommendation is do not give up your career for kids unless you have them and cannot afford to work or want to stay home. The traditional role of child care is only for the woman is a thing of the past. We all need to work together to improve the quality of life in our society.

Unfinished business is what everyone has in their life. We always have time to do better, try harder, change the things that are not working. American culture is one of innovation and acceptance. Kids do not know that there are stereotypical gender roles or careers that are unobtainable in other countries. We do not need to settle, women, minorities, immigrants, everyone in America is guaranteed a free comprehensive education that if utilized will lead them to furthering their education through trades, college, military, or work. As a society, we need to find ways to inspire personal ownership of education. We need to stop glorifying the media stunts and start celebrating the path to success. We never know when we will leave behind our life roles as influencers of a better future. Learning from the people around us and helping to make the world a better place is something that all decent people should strive for in their life.

Things to do in your classroom to help with the STEM engagement of all students. First find practical ways to get their hands dirty and make stuff. College will provide the full level of theoretical knowledge needed, secondary students need to have a safe space to build, iterate, fail, and try again. Never being told their idea is bad just encourages them to understand what to try next to make it better. Teachers are competing for attention with handheld technology. We need to make our show more engaging so the only reason they want their phone is to take pictures and videos to share with others the exciting things they are doing in class. Budget is limited in some schools, but I have found that even during the homebound, semester students could and did build Rube Goldberg machines with 100% participation. Make it exciting, experiment with students including personal failures, find the good in every project and encourage what to do next to make it better.

As a teacher, I can tell you that a little goes a long way when it comes to students. Experiences told me that if I got to know someone I would like them as a person. The hardest to like usually needed you the most. If you are not a teacher, there is a 2X10 theory that if you speak to a student for 2 minutes 10 school days in a role about things other than school and discipline their behavior will improve. This should be done at every job, school, community group you are in. If you are connected with the people you are around, they will take better care of you and you of them. Some of my hardest students to connect with due to personal trauma are some of my favorite of all time. You never know if you make a difference unless you see it. I have had "bad" kids help me break-up fights, share their lunch or treat with me, bring me gifts from their home country to thank me. Know me well enough to buy me my favorite soda and beef jerky for a Christmas gift. Parents sending gifts with their kids to their teachers is nice, but it melts my heart when a student brings me something they worked hard to get.

Little things are a key to building relationships of trust. The things I learn from students I connect with are all very educational even if it is them venting about gang and drug problems or students of a different race explaining to me how to work to encourage more minorities to work in STEM. Media is making it hard to teach because you feel like everything can be taken the wrong way and self-preservation is more important than treating others with respect. We know that our society is flawed when the first response when someone is in trouble is to take a video, TV shows dedicated to people failing at things they are good at, not providing safe places to learn about cultural differences in a way that is meaningful without judgment, these and many more are the reasons I teach and host exchange students.

I hope my efforts are exponential in growth, I hope Zach is a solid member of society, marries, has kids and they never face the issues he had as a kid. My exchange students are spreading world culture to me and taking some with themselves. Having dinner together almost every night with them was impactful and they took that example home to their home countries and now their families are getting to know each other as individuals. Being open to say I do not know, tell me more, can be exponential and should be acceptable. If you develop a respectful relationship with students, peers, and community they will trust when you ask the hard questions it is genuine and that you want to improve things for everyone. We do not need others to fail for us to succeed, life is not graded on a curve. Do better for yourself and others and we will all grow together in school, business, and society.

## ABOUT THE AUTHOR

**Kari Nathan** has BS in agricultural engineering technology and MSE: industrial technology curriculum and design along with secondary school administration certification. Being a lifelong learner, she has also obtained OSHA Trainer Certification, FAA 107 Drone pilot certification, NASA STEM training for educators, general contractor for two home builds, and mentors many students of a variety of socioeconomic and diverse backgrounds.

# 20 Five Amazing Women Making a Difference in STEM

*Michelle Segrest*
Navigate Content, Inc, Alabama, United States of America

**CONTENTS**



For more than 100 years, women have been forging the way for other women in science, technology, engineering, and mathematics (STEM) fields. One of the first to blaze the trail was Kate Gleason, who in 1918 was unanimously elected to the ASME as its first female member. The ASME had been around since 1880 so recognizing a female in the field took a while.

Fortunately, it didn't take long for others to follow.

Nancy D. Fitzroy was ASME's first woman president. Trailblazers like Yvonne C. Brill, Edith Clarke, Sally Ride, Mary Winston Jackson, Dorothy Lee, Lillian Moller Gilbreth, and many others made their marks in the profession of engineering at times when opportunities for women were limited.

But even with the growth of women in STEM fields, they are still underrepresented in engineering. According to the U.S. Department of Labor Statistics, women make up about 47% of the overall workforce, but only 14% of engineers are women.

While individual women are making their mark, some organizations are working hard to broaden their reach. The Girl Scouts reorganized its educational model and created its four pillars of content to include STEM programs. Organizations like Girls Who Code have programs with a mission to correct the gender imbalance by teaching computer programming to young girls. Technovation, as another example, empowers girls to become innovators and leaders through engineering and technology.

Here are the stories about five empowered women in STEM who are doing their part to make a difference.

DOI: 10.1201/9781003336495-23

## 20.1 INTRODUCE A GIRL TO ENGINEERING

At the 12th annual Siemens **"Introduce a Girl to Engineering"** event in West Chicago in March 2016, Jayne Beck showed a group of 100 girls what stuffed animals, cheesecake, music, perfume, airplanes, medical equipment, motors, and cell phones all had in common. It's easy, she explained. The designs all depend on the creativity and skills of engineers.

Beck, and other women like her, are breaking the male-dominated engineering stereotype. She is currently the Senior Engineer at Norton McMurray. She leads the Lean implementation and improves engineering best practices. Previously, she spent 30 years with Siemens, Inc., managing the team of engineers responsible for engineering-to-order motor control centers and switchboards and for designing mechanical assemblies for electrical equipment.

Her story, though, goes back even further.

"My dad was a plant manager of a very small company. He would take me with him to work on Saturdays and let me run manual brake presses and let me make things with scraps of sheet metal," Beck said. "I had girl toys, but also lots of boy toys. My parents didn't believe in stereotypes. I liked playing with Legos, building blocks, Lincoln Logs—toys more associated with boys, especially in the 1960s."

Her first aspiration was to become an architect. "I did well in science and math. A high school counselor told me engineering was the same as architecture," she said. "This is not really true, but I believed it and began to pursue engineering."

Women are still outnumbered 10 to 1 in the engineering field, something that Beck is still trying to change. "I think this stereotype has its roots when kids are very little," she explained. "There are stereotypes about the types of jobs men should have and the ones women should have. Most teachers and nurses are female, and most engineers are male."

Beck scored a summer internship doing drafting work for a company that hired her after she graduated from the University of Illinois, Urbana-Champaign. "I worked there for seven years and did a lot of mechanical design and also designed automated machinery," she said. "Then, I happened to find a job with another company in my area, Furnas Electric Company." Siemens acquired Furnas in 1996.

Being the only woman in a man's field has always been the norm for Beck. She hasn't let it stop her. A swimmer in high school, she draws on her competitive nature.

"I've always been a minority," she said. "As a manager, I was the only female on the staff for many years. In the earlier years, there were challenges related to being female. People just were not used to having women around. But it has gotten better and is not an issue anymore. There were always doubts about whether you can do the job because you are a female. I always felt like I had to work harder than the guys. I didn't want to fail so I was willing to put in long hours and extra effort to make sure I was successful."

During a 100th anniversary celebration for the Girl Scouts, a Siemens group was trying to encourage STEM (Science, Technology, Engineering, and Math) careers. There, Beck overheard a young girl telling her mother, "I'm going to be an engineer just like Daddy." Her mom said, 'Why would you do that?'" Beck admitted it was a bit discouraging.

Gender equality is important to Beck. So, it came as a shock when her daughter Katelyn made a startling comment at the tender age of 5.

> We were talking about homework, and she said, 'You know, Mom, girls can't do math.' I thought, how could my daughter truly believe that at such a young age? Someone had embedded that in her mind, and she believed it.

Studies show that girls do just as well in math and science as boys up to a certain age, but at some point, they change their path, Beck said.

> Sometimes the girls that do choose engineering do really well in high school, then find college more challenging. But college can be a quantum leap for everyone. People in general, and especially girls, don't seem to understand what engineers do. This plays into it not being a popular choice for girls. It's hard to break the pattern. Everyone goes to the doctor, so kids know what doctors do. People know what lawyers do. But not everyone understands what engineers do.

Katelyn Beck got the message loud and clear. She later became an architect and made a difference for her company. "It's a family joke now...the conversation about her not being good at math," Jayne Beck said. "Every once in a while, she jokingly says, 'You know, I can't do math.' But in the end, she found her passion and pursued it even though it wasn't the stereotype. She loves her job."

In addition to Katelyn, the Siemens **"Introduce a Girl to Engineering"** program has several graduates with positive stories to tell. Beck told the story of one woman who attended the event in 2006 and now writes computer programs for a company in Texas. Another guest makes helmets with interactive screens on the clear face guards for fighter pilots. "The possibilities are endless for girls in engineering," Beck said.

The Siemens program, offered to girls from fifth to twelfth grades, begins with a 30-minute interactive presentation that demonstrates the versatility and creativity of the engineering profession. That's where Beck shows the girls how things such as cheesecake, perfume, and stuffed animals require engineers to manufacture them.

"I explain the roles engineers have in cheesecake, for example," Beck said.

> I've been to a cheesecake factory in the Chicago suburbs where they have incredible automation with cheesecakes flying down the assembly line and giant industrial-sized mixers. I explain how it takes engineers to make perfume because they may help to create the chemical formulation and the processes for manufacturing. Engineers are involved in developing the formulas for the fabric and stuffing and little plastic eyes for stuffed animals. Engineers design the mold to make those eyes. The point is...engineers play a role in anything that is man-made.

Beck asks the girls key questions such as, "Why are you letting the boys have all the good jobs?"

"I hold up my cell phone and ask if they care about this," she said. "I let them know they have cell phones because of engineers."

Beck gives more examples about the diversity of engineering jobs.

"The perfect training for a patent lawyer would be an engineering degree," she said.

> In the medical field, Siemens Healthcare hires engineers to design all sorts of medical equipment and devices that improve people's lives. Most astronauts are engineers. A lot of girls don't realize there is a huge market for engineers in sales. If they are doing a great job selling Girl Scout cookies, maybe they could go into technical sales as an engineer.

The last part of the presentation includes 10 questions.

1. Do you want to make the world a better place to live?
2. Do you have a major imagination?
3. Are you good with nitty-gritty details?
4. Do you consider yourself an inventor?
5. Do you like being a problem solver?
6. Do you enjoy group projects?
7. Are you totally into drawing, sketching, or building things?
8. Do you like to show people how things work?
9. Did you get good grades in science or math last year?
10. Are your writing skills pretty good?

If the answer to at least five of these questions is "Yes," then a career in engineering may be a good idea.

Following Beck's introductory presentation, attendees are treated to a tour of the Siemens West Chicago facility. At each of the 12 stops, there is an explanation about how engineering contributes to that process.

The interactive portion of the program includes building a tower with dried spaghetti and marshmallows, building a wind turbine that actually generates energy, building a boat from a 6 x 6-in. piece of tin foil, and a paper-airplane contest.

Beck wears the banner for women in engineering proudly.

"I would like to think that people see me now as an engineer and not a female engineer," she said. "It's still rare to find a female engineer, but I don't think this is a negative thing. I'm proud of my career and grateful for the opportunities I've had."

Still, stereotypes and misunderstandings can get in the way of communicating the value of engineering to a younger generation.

"Engineers have gotten a bad rap in general through the years as being nerds," she said.

> I think that is changing. There is more respect in the schools for having academic achievements over sports. Now there are math teams and robotics teams, so I think our culture in general is improving with regard to technical jobs. We need to encourage boys and girls to consider STEM careers. It's a significant need in our country.

## 20.2 WOMEN SOLVE WATER INDUSTRY CHALLENGES

For Indar's Water Engineering Director Elena Rodríguez, expanding her knowledge base every day helps her tackle complex water projects worldwide.

Rodríguez considers herself a full-time student. Even with more than a quarter of a century of experience in turbomachinery and hydraulics engineering, she never stops learning.

"Being a pump engineer is a very creative job," said Rodríguez.

> Different projects require totally different solutions, so I always have the opportunity of developing new ideas and designs and at the same time challenging myself. You feel like you are always learning something, and that is incredible for me. Indar is a custom-oriented company with a strong R&D and engineering capabilities facing innovative projects under challenging conditions. As a pump engineer, that is a gift!

Rodríguez is based in Spain and finds herself working various roles on multiple projects worldwide. As a water engineering director, Rodríguez works on projects from the very beginning, when they are only conceptual ideas.

"Being able to solve the existing needs in terms of water is really exciting," she said. "I am thinking about the best solution for the project needs, participating in the design, tracking, and interacting during the complete process until the pumps are installed."

The Lake Mead Low-Lift Pumping Station Project in Nevada and the Carlsbad Desalination Plant Dilution Pumping Station in California are two of her notable projects.

Lake Mead is the largest freshwater reservoir in the United States in terms of maximum water capacity. It was recently at its lowest water level since the lake was first filled during construction of the Hoover Dam in the 1930s. A new pumping station (L3PS) requiring submersible pumps is currently in the works. Lake Mead L3PS substantially increases the reliability of Southern Nevada's water delivery system.

The Claude "Bud" Lewis Carlsbad Desalination Plant is the largest, most technologically advanced, and energy-efficient seawater desalination plant in the U.S. With the decommissioning of Encina Power Station, the Carlsbad Desalination Plant is modernizing the existing intake facilities to provide additional environmental enhancements to protect and preserve the marine environment complying with the regulations in the California State Water Board's Ocean Plan Amendment.

The new pumping station uses the biggest fish-friendly submersible pumps in the world. The innovative solution of Dilution PS, for Carlsbad Desalination Plant, will allow protection and preservation of the lagoon so that the community can enjoy its recreational and marine resources now and for future generations.

Rodríguez has diverse experience. She considers her most vital areas of expertise in hydraulic engineering, pump design, and pumping system definition. In 2008, 10 years after completing her master's in mechanical engineering, she decided to further enhance her education and training. She earned an additional degree in civil engineering with a specialty in public works.

"Having a complete understanding and knowledge of the systems allows me to define the best solution for existing hydraulic needs," she said.

Her desire for learning and her aspirations of becoming an engineer began when she was a little girl.

"From a very young age I have always liked science, physics, and mathematics, more in practical ways than theoretical, so becoming an engineer was simply the right choice for me," she said.

> Becoming a pump engineer was nearly by accident, or to be more exact, by pure coincidence. It was 1997 while I was still a student working toward my master's in mechanical engineering. Summer was around the corner, and I wanted to start with something in the industry. One of my core subjects was hydraulic machinery: pumps and turbine. Taking into account that there is a pump company in my hometown (Zarautz, Spain – Bombas Itur now part of KSB), I asked for a summer internship.

Upon completion of her degree, she was offered the opportunity of completing her master's final project on CFD analysis of an existing centrifugal pump.

"This was a very challenging project as it was the first one in the company and one of the first ones of this nature in Spain," she explained. "I successfully finished the project with honors in 1999. This experience was the spark for my passion with pumps."

After a few months in a technological center working on CFD codes, in July 2000, Rodríguez began working with Indar Submersible Pumps and motors as a pump designer.

As with most females in a male-dominated profession, there were additional challenges to tackle, but she never considered being a woman engineer an insurmountable obstacle.

"Professionally, I have never considered that being a woman means a limitation, even if sometimes you feel like pushing yourself out of your comfort zone," she explained. "Personally, with my job being able to inspire new generations of women to become pump engineers and transmitting my passion for this market, I consider this among my greatest accomplishments."

Breaking stereotypes is important to Rodríguez.

"There were not many women in the University," she remembered.

> In fact, I think there were just five or six of us in the Master of Mechanical Engineering program. That was already showing the tendency in the companies. Even now, we are a minority in many engineering fields. To me, it is quite normal being the only woman engineer in many of my daily meetings. I have no problem with the situation. I never think about this being a handicap. However, at the beginning, it was not easy. The combination of being young plus being a woman was a very challenging and difficult scenario for a woman engineer.

The obstacles are there, but not impossible to overcome, she said.

"In my opinion, maybe there is initial distrust for being a woman engineer," she explained.

> But if you trust your abilities and support yourself, it is easy to break the barrier. It is important not to fool yourself. Not everything is easy, but you have the power inside, and you must always believe in yourself—always.

When she is not working, Rodríguez loves to travel and spend time with family, especially her niece who one day wants to be an engineer.

> The different projects I am involved with take me to different parts of the world. I like to think that with my job I contribute to community well-being. Water is a limited natural resource, fundamental for life and for all economic activities.

Rodríguez said she can't imagine doing anything other than being a pump engineer.

"Now, it's just part of my essence, and honestly, I love it!" she said.

## 20.3 IGNORING THE GLASS CEILING

Rendela Wenzel approaches each maintenance issue like a CSI investigator. "I love to understand how things work and explore the mystery behind equipment failures," she said. "The main difference between a police detective and me is that I investigate and fix machines, not people."

A global consultant engineer, Wenzel uses every tool in her toolbox. However, her most effective tools are experience, analytical thinking, training, and a belief that every option should be exhausted.

"Let's say we have a bearing issue. I can look at the peaks and tell you how many side bands there are and what frequencies are off," said Wenzel, who specializes in maintenance-and-reliability engineering for Eli Lilly, headquartered in Indianapolis.

> I can pinpoint the approximate amount of life left in the bearing. I can tell you if the lubrication was adequate. We can use thermography to look at the heat signature. Like a CSI investigator, I go in and examine all the details, then troubleshoot a problem from the inside out.

Her first step is to examine the body. "You can't do an investigation without the body. Instead of a body, I have a piece of equipment," she said.

> The chalk line is drawn, and they bring me the part. They may bring the bearing or the pulley or the whole pump. We have an area in predictive maintenance where they will go in and cut those apart. We then write tech reports and give it to management. It's like a police investigation, and I get to do all the reporting around it.

Wenzel's enthusiasm for diagnosing and repairing machines started at an early age. Born into a working class, blue-collar family in Terre Haute, Indiana, Wenzel earned a mechanical engineering degree from Purdue University in West Lafayette, Indiana, and became the first person in her family to attend and graduate from college. Her father worked on a manufacturing line at Pillsbury, and her mother worked from home as a cake decorator.

"Growing up without a whole lot, I learned to make do and get creative," Wenzel said. "I spent a lot of time working with my hands and fixing things. I would fix bicycles, tinker with cars, and I would build things from wood."

Her original aspirations were to fly airplanes and become an astronaut. That would require military service and an engineering education.

"I like the theory side of engineering, but I also like working with my hands. It helps me to understand things better," she said.

> What I love about what I do with reliability engineering is I can go out and work with the crafts. Then, I can write the report. I interface with higher levels of management, craftsmen, and engineering personnel, and travel to different sites and help them solve problems.

She gained early management experience as a captain and Quarter Master Company Commander in the U. S. Army and as a maintenance engineer and supervisor at Chrysler and International Truck and Engine Corp.

Now, Wenzel designs and implements programs then facilitate the reliability discussions and onsite failure analysis for the company's 21 manufacturing sites in 13 countries.

Being a female in a male-dominated industry was tough at first, she said, but she adapted quickly.

"There's been some good, some bad and some ugly," she said. "My first experience out of college was with Chrysler in 1997. Entering a leadership role in a foundry was intimidating. There was one forklift driver. We called him "Tramp." He saw me and said 'Sweetheart, are you lost?' I said, 'No, I'm your boss.' He was an older gentleman and had never had a woman supervisor in 40 years in the industry. He and I became fast friends, and he was one of our best employees."

Wenzel experienced the metaphoric glass ceiling quickly but didn't let it stop her. "There can be a disparity, especially among engineers, as you rise in the ranks," she said.

> The more experience I get and the older I get, I find I'm held to a higher standard. It's just that much tougher, but it is an adjustment that you make with time. It becomes a part of your personality and a part of who you are. I believe strongly that I should not be given a position because I am a female. Give me the position because I'm qualified and the best person for the position.

A common practice with many motor manufacturers, Chrysler had a philosophy of having engineers spend a few years on the floor in management to learn how the business works from a grass-roots level, Wenzel recalled. "This helped me to learn how to manage people and also manage assets," she said.

At International Truck and Engine, Wenzel wrote her own job description as the company embraced an environment of reliability and predictive maintenance. "They told me to find out what these technologies are and then bring them back to the company," Wenzel said.

Her title of mechanical engineer was transitioned to reliability engineer. She implemented an oil-analysis predictability training program that included vibration, oil, and thermography.

"At International, it was in a union environment, so it had a different spin," Wenzel said.

> I had to learn how to troubleshoot equipment without touching it. Only the craftsmen could touch the equipment. I had to learn how to explain to them how to fix something instead of touching it and fixing it myself like I had always done. There were times when they would ask 'What do you mean?' I would explain how the item worked, then ask them to tell me from a mechanical standpoint how to fix it. We would then create a hybrid approach. It was very interesting to learn how to communicate it instead of doing it. It was like working with one arm. But I got very good at it, which has helped me in my current position because I sometimes have to help fix problems all over the world over the phone.

Wenzel works with a variety of pharmaceutical-grade equipment, including bulk pharma pumps, tanks, agitators, vacuum dryers, chromatography columns, vacuum units, and buffer systems.

"There is a lot going on that affects the chemistry inside that tank," she said.

> I may not understand the chemistry inside that tank, but I understand the facets of the mechanisms needed to deliver it. Everything is clean and well maintained. We are very mindful of patient safety. With everything you do, you have to remember that this could be for your husband, or for your wife, or for your child. That is something culturally that is on the minds of everyone.

Wenzel's experience spans many roles in a variety of industries. Each one provides the opportunity to learn.

"As a senior-level engineer, you have to be able to influence without direct reports. I must be able to operate as a manager, but without the overhead responsibility of managing those people," she explained.

> That also has its limitations – to influence those people technically without being in charge of them. In a global world, you must be able to influence and get the agenda across, and implement programs at each site, even when everyone is so culturally different.

## 20.4 ENGINEERING EFFECTIVE RELIABILITY STRATEGIES

As a young girl, Gina Hutto Kittle would sit in the garage with her father and grandfather and study their every movement. She watched her father – a mechanic – fix anything that the neighbors needed repaired.

Her grandfather, Owen Ramsey, worked with Red Stone Arsenal in Huntsville, Alabama, where he was part of the core group that launched the first missile into space. Kittle would play with his drafting tools and ask hundreds of questions about how things worked and how to fix them when they broke.

Even though she didn't really know what engineers did, she knew she wanted to be one.

"My grandfather passed away when I was in the fifth grade, but I remember seeing the newspaper articles about his involvement with the missile," Kittle said.

> I found what he did so interesting. He worked on cars and would show me what he was doing. I learned about all the tools and how to use them. This was manual machinery, and just being around it inspired me. Even though I didn't really know the definition of an engineer, I just kept telling everyone, 'I'm going to be an engineer.'

Kittle now has a mechanical engineering degree and is the ANH Global Reliability Leader at Cargill. She spent more than 22 years as the program manager for The Timken Company's manufacturing advancement program, product and process development.

Kittle also found time to serve as the treasurer, then secretary, and then member services director of the executive committee for the Society for Maintenance and Reliability Professionals (SMRP), Atlanta.

"Within SMRP, I have been able to work with some really great people," Kittle said. "It has given me a platform to learn and share knowledge about the reliability industry."

Kittle remembers being a bit overwhelmed by her first SMRP conference 20 years ago.

"I didn't even know half of the terminology back then, and I was blown away by the knowledge that was in the room," she said.

> The next year I began to get engaged with the committees and worked my way to conference chair. This was a huge experience seeing thousands of people come together to improve maintenance and reliability. It's amazing. I'm making bearings, but you may work with someone who is making ice cream, or someone making insulation or producing fruit or making pistachios. But we all have the same issues.

Kittle is aware that she is a woman in a male-dominated profession, but she doesn't let that drive her. "Especially in today's world, my goal is to be known as a mechanical engineer with the right credibility and the right certifications," she explained. "I want to be held up on the merit of my 20-plus years of experience. It shouldn't matter whether I'm a female. However, it's important to bring diversity to how groups like SMRP interact."

Even with this positive attitude, she can't ignore that diversity has not always come easy in business, and particularly in manufacturing. "It has been a struggle," she admitted.

> When I first started out in the mid-to-late 1990s, there were contractors that didn't want to deal with me. They would walk right by me and shake my boss' hand even though I was in charge of the project. There are times when women in our field must overcome things like this, but at the same time, you can't let it weigh you down.

Kittle offers advice for all young engineers – male and female.

"As women, we have to protect our image a little more than men do," she explained,

> from little things like how we dress to big things like being sure that our voice is heard. I would advise any young engineer to put your time and effort into making sure that the information you have is correct. If you say something, be sure you've done the research and are certain this is the way to go. Once you lose credibility, it's hard to get it back.

Kittle said she will never forget the lessons learned from all her mentors and experiences.

"Growing up, I had the opportunity to attend an Engineer-for-a-Day event at a local company in Tennessee," she said. "That sealed the deal for me wanting to be an engineer. Once I went to college, I started in the co-op program. Many of my projects on my assignments involved working with the maintenance technicians. After my first year of that assignment, I was hooked and knew this was the group I wanted to work with and with whom I felt the most at home.

"Being able to solve a problem and make something work again was what drew me in at first, but then trying to figure out how to make it even better was the next natural step."

Striving to improve is always a part of what drives her. "I love this field and have lived it practically my whole life," Kittle said. "My grandfather always said to always try to be your best. You'll never reach perfection, but if you don't at least strive for perfection, you'll never even get close."

## 20.5 UNDERSTANDING THE MECHANICS OF MACHINES

Rebekah Macko approaches every task with one simple philosophy in mind: It's more important to get it done right than to get it done right now. Her father taught her that lesson early in life. An organic geochemist, his approach to all things revolved around a methodical process, the success of which he passed down to his daughter.

"I can remember spending a lot of late nights in the lab with my Dad," Macko said.

> In fact, he likes to tell the story that he took me to the lab before taking me home from the hospital when I was born. We would take apart something, and he would show me how it works. I was always encouraged to look at the world around me, appreciate science, and look at the details of things. When I was really young, I thought I wanted to be a biologist and work with bugs or work in the rainforest.

As she grew older, around the age of 10, Macko began to realize that she had an aptitude for understanding the mechanics of machines. "Science was really romantic and fascinating to me," she said. "I thought it was so amazing what people could do with all these creative tools and instruments. That's when I really began to think about being an engineer."

The interest stuck. Every day, the senior fluid-systems engineer with Geiger Pump & Equipment in Aston, Pennsylvania applies the lessons learned from early experiences in the lab with her father.

"I remember one time replacing the coil on a gas chromatograph," she explained.

> It's a pretty simple operation with a big narrow coil of wire. I had seen the machine a bunch of times, but I didn't know what was inside, or how it worked, or even what it did. My dad explained what the parts were, and what was behind the 'secret door.' I loved learning how it works and how it tells us useful things. Now, I always want to know what every little mechanism is and the parts behind it. It's fascinating to me to realize that someone figured all this out. Nothing just happens. There are many different components, and pieces, and parts needed to put together every puzzle.

Macko uses her curiosity and mechanical gifts every day with Geiger. As the team leader of the packaged-systems group, she does everything from taking customer calls to troubleshooting problems and testing equipment under construction. She also works with the quality-assurance program and spearheaded the company's successful effort to achieve ISO9001 certification.

"Rebekah is a pleasure to have at Geiger," said the company's president, Henry Peck.

> Rebekah is a champion for keeping our systems group on task in delivering the highest-quality products and services. When our team has an engineering challenge, whether technical or logistical, Rebekah is a go-to for helping to find a solution. She is a role-model employee for our favorite Geiger values – continuous improvement and collaborating as a team.

Geiger's overall base business is in industrial pump distribution, parts, assembly, and repair. Many years ago, the company (then Smith-Koch), responding to market demands for the increased quality and simplified logistics of factory-assembled fluid-handling systems, started building packaged fuel-oil systems. Since then, its packaged systems business has expanded into custom-engineered products, building increasingly complex systems specifically to meet customer needs.

It requires careful attention to individual engineering requirements. Macko's main responsibility is to ensure that the company meets customer requirements and delivers a complete packaged working system.

She uses her instinctive and genetic methodical system to coordinate the efforts for her seven-member team. Her process begins by compiling as much information as possible.

"For example, a customer may be replacing a particular bad actor or looking to optimize a piece of pumping equipment," Macko explained.

> Sometimes people will send a grainy chart or a blurry picture, and this can actually be helpful, but you need a firm footing. Information is the key. Then, I look at the details of what has been done before and try to decide if this is the way we want to continue. I'm of the younger generation of engineers, so this gives me the freedom to make new suggestions, or to try a new technology. For example, personnel may be controlling the pump with pressure, but perhaps they should be using a different variable, like level or viscosity, in their control scheme.

The next step is selecting the right equipment for the job.

"People love to oversize their pumps," Macko said.

> They feel more comfortable with a fudge factor. There is always uncertainty in the real world, but with enough information you can pick the right fudge factor, and that can make all the difference. Sometimes it takes courage to try something new, but I'm a firm believer in designing your system with an investment in some flexibility. You just need a solid Plan B. Every plant is unique, and every installation is unique. If the information you begin with is correct, then newer technology like variable-speed drives and proper implementation can help a lot.

Before working with Geiger, Macko worked in the public works department for the city of Charlottesville, Virginia. She inspected catch basins and mapped stormwater systems, literally working in the middle of the highway. This experience helped her understand the connection between the theory and the field.

"It's so important to have a feeling for how things work, and this is developed through experience—taking the pump apart, putting it back together, getting to know the machines, feeling how they work," she explained. "But there is engineering and math behind every piece of it. And the two agree. The two go together. If they don't, you need to adjust one or the other."

Like solving a puzzle each day, finding creative solutions is a driving force for Macko. "It's a matter of looking at resources available beyond just the pump," she said.

> It's important to look at the whole system, use some pump math and graphing, and then put together the theory. I also find a lot of value in going to the site, meeting with the customer, putting my hands on the equipment. It takes extra time, but it's worth it. And it goes back to getting all the information you possibly can. You can't always do this from the other end of a telephone.

Once she has as much information as possible, the real work begins. "When I have all the information, then I can break down the problem into testable hypotheses," she said. "We can determine answers to questions like, 'What, beyond the obvious, could be contributing to the issue?' and 'What do we know about recent changes in the system?' Nothing is in isolation.

> Then, I think about it and try to reframe the question. If it's something that's really stumping me, I try to describe it in three ways – with words, with diagrams, and with math. The hardest, the way that makes me think most closely about the problem, is usually in the math. I may go back to my textbook and think about the principles at work. Or I may draw a diagram and discuss it with someone with a different knowledge base than me. I try to get a different angle, and think about it in a granular way so I can process it differently. I come up with lists of possible causes and try to figure out how to test each one.

Even when she is not working, Macko's engineering side creeps in. "It's just who I am," she said. "And I wear the 'nerd' sign like a badge of honor."

For example, in her ceramics class, while the other students were creating pots and bowls, Macko enjoyed making pinhole cameras. "For me, pottery is a nice marriage of engineering, planning, and creativity," she said.

> You draw things, sketch them out, and make plans. I would make the pinhole cameras out of clay, throwing the body of the camera on the potter's wheel, make a lid, and plan for the pinhole, shutter, and holder for film to adjust for focal length. Exposing polaroid film through the pinhole you can get all kinds of wild images. This is how an engineer does pottery, I guess. I always had the best measuring tools. Others would say, 'I really love the visceral feel of the clay,' and there I was with my protractor.

Determined in personal and professional challenges, Macko draws inspiration from the famous Marge Piercy poem, "To Be of Use," which describes the importance of putting your shoulder into your work and getting things done right. "We are not doing philosophy here," Macko said.

> We have to get it right. We can't take short cuts, because if we make mistakes, it will be seen. I believe that if you are going to do something, do it right. This philosophy rolls into my everyday life and with everything I do. My son doesn't have to become an engineer, but I hope he grows up to be determined and creative and enthusiastic about the world around him. I hope that he can always find something interesting in everything in his corner of the world.

## ABOUT THE AUTHOR

**Michelle Segrest** is the president of Navigate Content, Inc. and has been a professional journalist for more than three decades. She has specialized in creating content for the municipal and industrial processing industries since 2008. Check out her 3-volume book series on Modern Manufacturing.

# Index

L



M



N



O



P



Q



R



S



T

**U**



**V**

Printed in the United States
by Baker & Taylor Publisher Services